DOUGLAS HAIG
AS I KNEW HIM

To / The Rev: G. S. Duncan,
      My Chaplain 1916 to the End –
      in all gratitude,
                    D. Haig. F.M.
                    Xmas 1918.

DOUGLAS HAIG
AS I KNEW HIM

GEORGE S DUNCAN

Pen & Sword
MILITARY

First published in Great Britain in 1966 by George Allen & Unwin Ltd
and reprinted in this format in 2015 by
PEN & SWORD MILITARY
An imprint of
Pen & Sword Books Ltd
47 Church Street, Barnsley
South Yorkshire
S70 2AS

Copyright © The Douglas Haig Fellowship (Registered as an educational charity
in England & Wales No: 1150833 & Scotland SC044690), 2015

ISBN 978 1 47382 770 7

Printed and bound in England
By CPI Group (UK) Ltd, Croydon, CR0 4YY

Pen & Sword Books Ltd incorporates the Imprints of Pen & Sword Aviation,
Pen & Sword Family History, Pen & Sword Maritime, Pen & Sword Military,
Pen & Sword Discovery, Pen & Sword Politics, Pen & Sword Atlas,
Pen & Sword Archaeology, Wharncliffe Local History, Leo Cooper,
Wharncliffe True Crime, Wharncliffe Transport, Pen & Sword Select,
Pen & Sword Military Classics, The Praetorian Press, Claymore Press,
Remember When, Seaforth Publishing and Frontline Publishing

For a complete list of Pen & Sword titles please contact
PEN & SWORD BOOKS LIMITED
47 Church Street, Barnsley, South Yorkshire, S70 2AS, England
E-mail: enquiries@pen-and-sword.co.uk
Website: www.pen-and-sword.co.uk

INTRODUCTION

This book was originally published in 1966, shortly after the author, Reverend George Duncan, had died.

The climate of the time was not particularly propitious for what is a defensive book on Haig, as a man of grounded faith and as a military leader in the Great War. The new generation of post-Second World War 'baby boomers' were soon to be in revolt against their parents' generation (perhaps best exemplified by the events of 1968), against practically all things military, against the Establishment and organised religion. Possibly, this has led to it being dismissed by many students of Haig, the British general about whom probably more printers' ink has been used than any other British military leader (with the possible exception of the Duke of Wellington). Duncan, being a fully paid-up member of the Establishment, both academic and ecclesiastical, who was Haig's spiritual guru, could not possibly have anything useful to say about the man whom many authors were describing as 'Butcher Haig'. It would be a polemical, apologist account of no significant historical value other than as a curiosity and as a coda to the work of the revisionist historian John Terraine, whose *Haig: The Educated Soldier* (1963) was – and in truth is still – seen by many as defending the indefensible. Indeed, Duncan acknowledged the assistance given to him by Terraine in writing his book. It was the era of the hugely successful *Oh, What a Lovely War!* (first performed in 1963), following on from Alan Clark's popular *The Donkeys* (1961) and Leon Wolff's 1959 indictment of the Third Battle of Ypres (Passchendaele), *In Flanders Fields*.

A subliminal subtext to the subsequent historiography on Haig in particular, and the performance of the BEF in general, is that Duncan could not have much of use to contribute to the argument. In the edited version of Duncan's diaries by Gerard de Groot, 'The Reverend George S Duncan at GHQ, 1916-1918', in the Army Records Society's volume *Military Miscellany I* (1996), he is fairly damning about Duncan's ability to be any sort of impartial observer, although his wartime experience and his mingling with the great and the good at GHQ – and the fact that he was a chaplain who worked extensively with junior officers and other ranks – provided value to the diaries for their insights. However, there are some fairly dismissive passages: in 1916 Duncan 'looked more like a young, suitably eccentric university professor than a potential Messiah [this latter refers to Haig's admiration of him as a preacher]'. I am not totally convinced that contemporary photographs support that view point; I leave it to readers to judge. 'Orphaned at an early age, he quite possibly found a father-figure in Haig [there again, possibly not], who he likened to "an old Homeric hero". He gave Haig unquestioning loyalty, which bordered on adoration.' De Groot is no great admirer of Haig and certainly felt that Duncan provided the perfect adjunct for the great man's rekindled religious fervour and use of the Divinity as a moral support to see him through the war and to reinforce the validity of his decisions: précised, God is on our side and all will be well. Denis Winter, in his *Haig's Command* (1991), goes further: 'Haig's denomination seems almost to have been chosen as a result of a particular preacher's good looks [so much for the 'eccentric university professor'], youthful energy and simple sermons.' It is difficult to know quite where to begin in criticising that statement.

It is as well to outline the career of this cleric who, undoubtedly, was a support to Haig during the war. George Simpson Duncan was born in 1884 in Forfar, the son a tailor's cutter. Unusually for someone of such a modest financial background, he went to Edinburgh University and gained a first class degree in Classics (1906). This was followed by a scholarship to Trinity College, Cambridge and further studies at St Andrews, Marburg, Jena and Heidelberg – indeed, in later life Duncan became well known for his desire to improve relations with the Lutheran and Reformed Churches, doubtless originally kindled by his time in Germany at some of its finest Protestant scriptural and theological departments.

In 1915, Duncan was ordained in the Church of Scotland (the denomination in which Haig had been raised) and volunteered for service as a military chaplain. Based at St Omer, the then location of GHQ, he made an impact on Haig when the latter attended a service conducted by Duncan soon after his appointment as the new commander in chief in mid-December. His wartime service will be briefly covered later; but for the moment it is important to establish what sort of credentials he might have to write a considered examination of Haig.

In 1919 he became Professor of Biblical Criticism at St Andrews University, quite a notable achievement for someone of his age; he held the Chair for the rest of his working life. He became Principal of St Mary's College in the university in 1940 and its vice chancellor (1952-53); in 1949 he was Moderator of the General Assembly of the Church of Scotland and vice president of the British Council of Churches (1950-52). He was a major contributor to the translation and production of the New Testament section (1961) of *The New English Bible* and proved himself, with his common touch, to be an adept and accessible exponent of it on television after its publication. In addition to all of this he worked with Lutheran and reformed churches on the continent, a contribution whose value was acknowledged by a clutch of honorary degrees and honorary professorships. Duncan was acknowledged as an effective teacher, with a committed, natural and easy relationship with several generations of students and candidates for ministry. He died in April 1965 – very shortly after he had completed the manuscript of the book.

The point of this rather lengthy biographical note is to show that by the time he came to write this work, he was a man of some practical experience – admittedly not in military matters; but certainly covering a wide range of challenges, from that of executive administration to incisive scholarship and pastoral and spiritual care both at a national level, and with and for numerous individuals.

In any examination of the contents of a book, I would suggest that it is always useful to see how much it keeps to the title; and here Duncan has laid down his theme quite specifically: *Douglas Haig As I Knew Him*. In his preface he makes clear what his objective is: 'I have written primarily for the general reader, who may wish to have a picture of Haig from one who saw much of him at close quarters during those years of war. But I trust that the book may also be of service to the serious student, helping perhaps by a statement here, a testimony there, to shed light on the character and achievements of this great soldier.' The preceding paragraph to this shows why he felt moved to write the book after the war had been over for so many years and long after Haig had died: he notes that many writers 'have presented to the public a portrait of Haig which is so distorted as to be essentially false. The time would seem to have come for a fresh appraisal of his greatness both as a man and as a commander...'

So we have been clearly signalled about what to expect. If further evidence is needed, the quote he selects from Tacitus (in his introductory comments to his *Life of Agricola*) to precede his introduction says it all: '... and when a man of distinction arises, the sinister tendency (found alike in states great and small) to ignore or disparage true worth becomes powerless in his presence – his greatness lifts him high beyond its reach.'

The book is both well written and well crafted, which I think can be accepted by all, regardless of their views on its subject. He sets the scene – including some interesting comments on the impact of two world wars, in particular the impact of the second on perceptions of the first. This might be stating the blindingly obvious, but in historiographical terms this reality is all too apparent in much of the writing on the Great War and on generalship – Haig's in particular, perhaps – with all that knowledge of the cataclysm of 1939-1945 and the events that led up to it brings to bear on studies of the First World War.

This first chapter shows, to my mind, one of the most challenging aspects of the book. Although he does not necessarily answer all the issues that he raises – how could he in so short a book, which in reality is something of an extended essay – he does raise them. When re-reading it in preparation for writing this introduction, I was struck by how frequently the points that he mentions would provide numerous topics for full-length research papers. In fact anybody seeking to write an assessment of Haig, both as a person and as a military commander, would be well advised to use this as a form of check list, to see if all the relevant points have both been covered and covered in appropriate depth.

The value to his writing lies in his experience in wartime. He is quite frank about his qualifications in writing what he does. He acknowledges his complete lack of military background: '...I knew next to nothing about army life when in 1915 I was selected for a commission ... I never had a pastoral charge...'. On arrival in France, in September 1915, he was posted to St Omer, French's GHQ, and worked amongst the troops based there and in four near-by hospitals. He first encountered Haig when the latter turned up, unannounced, on the first Sunday in January to the modest room (and even more modest congregation, it seems) used for the Presbyterian service.

Thereafter Haig was a regular, but intermittent, attendee at his services (to his surprise he was transferred with the rest of GHQ when it was moved to Montreuil in March 1916); intermittent because Haig was not always in GHQ and because sometimes he attended Anglican services. It would also be quite wrong to think of him as 'Haig's chaplain', a full-time occupation: far from it. However, it is quite clear that he regarded the situation as a major responsibility; and it is equally clear that Haig regarded him as a major source of spiritual comfort: no arguments there. In addition to his time at GHQ, he was also invited to follow Haig on occasions when he moved to an Advanced GHQ during a major offensive; at the same time he took very seriously his pastoral responsibilities to the men serving at GHQ – and it was almost exclusively a male preserve until late into the war – and also worked as a chaplain extensively in casualty clearing stations: he had plenty of contact with the ordinary soldier.

The first section of the book, therefore, introduces us to how Duncan came to know Haig; something of Duncan's duties; his initial reactions to Haig as the person who was commander in chief; his own life at GHQ and how his opinions developed over the time that they were together during the war – he was, by and large, a silent but privileged observer, often being invited to meals with the great man. These opinions are nuanced by what was written about Haig, generally by those who knew him relatively well, in the years after the war. His short chapter on life at GHQ is particularly useful.

The second section, entitled 'Historical' – by far the longest part of the book – examines his relationship with Haig for the rest of his life, obviously with the emphasis on the war years; though it is worth bearing in mind that Haig was Rector of St Andrews (1916 – 1919) and then its Chancellor from 1922 until his death and therefore had a professional – as well as a personal – reason to maintain his contacts with Duncan post war. The chapters in this section broadly split the war years into the dominant events of the war; therefore there is a commentary on what was happening, how he perceived events to impact on Haig and his own doings in relationship to the chief in each period. In the process of doing this it is evident that he was regarded by other members of the staff, particularly those who might be regarded as Haig's household staff, as discrete and intelligent.

The final section is entitled 'Personal'. Although the second section has much to offer the reader, not least because of some of the issues and questions that it raises, it is this final section which is possibly the most useful from the perspective of the personal judgements made by Duncan about Haig. The chapter headings are instructive: 'The Man we Knew; The Man of Character; The Man of Faith'. In particular I would place great importance on the chapter on Haig as a man of faith. It is rather less than six pages long; it sums up the reflections of a highly experienced – and much respected – expert, for want of a better word, in this area of a person's life. It seems that too many modern studies of Haig simply do not understand – or want to understand – the place of faith as part of the make-up of the character and perspectives of an individual; especially an individual who was brought up in the norms of late Victorian Britain. On this matter Duncan summed up Haig's approach: 'We ought to think of him primarily as a man with a simply personal religion, a religion which profoundly affected his whole outlook on life and his relations with his fellows.'

Duncan's book never aspired to be anything more than what the title suggested it would be; an examination of the life of Haig in the light of Duncan's experience of him and his understanding of the events of that particularly stressful time. It is in no sense a biography, let alone a military analysis of his capabilities. On the other hand, had the book not been published it would have deprived students interested in Haig of an insight into a man of wisdom, who was a semi-detached, intelligent observer. Not to mention a commander in chief in the midst of the greatest war the nation had know up to that point, and who was under considerable stress for the three long and tortuous years that he held this awesome responsibility.

FOREWORD

BY SIR ARTHUR BRYANT

Many who have passed or are approaching the psalmist's span served in youth under the command of Douglas Haig. It has long been the fashion to denigrate him as an unimaginative commander whose preference for frontal attack in the face of modern automatic weapons caused casualties that a more intelligent soldier could have avoided. One modern writer, embroidering on this theme, went so far as to suggest that the force which under Haig withstood and broke the strongest and most superbly organised military machine in history was an army of lions commanded by donkeys.

Yet, despite the appalling casualties of those years, few serving under Haig regarded him at the time as other than a soldier of the highest competence and the one man capable of commanding what had grown into by far the largest army Britain had ever put into the field. Raised by voluntary recruitment from an untrained civilian population and pitted against the most formidable army in the world at a time when Russia and Italy were on the point of collapse and French morale and elan had been shattered by the casualties of Verdun and Nivelle's disastrous offensive, it far exceeded in numbers the largest expeditionary force commanded by Roberts before or Montgomery afterwards; the only British military commanders who have borne a weight of overall responsibility comparable to Haig's were Marlborough and Wellington, and in different and global context, Alanbrooke. The great Liberal Secretary-of-State-for-War, Haldane, described him as 'the most highly equipped thinker in the British Army, . . . the only military leader we possess with the power of thinking, which the enemy possesses in a highly developed form'. At first Ypres, before he was given the supreme command, his troops saved the Channel Ports in one of the most desperate and crucial holding-actions in our history, eliciting from the Kaiser the tribute that 'the British First Army Corps under Douglas Haig is the finest in the world'.

Without confidence in its commander-in-chief no army could have sustained in two successive years the losses of the Somme and Paschendaele—both sacrificial offensives undertaken to relieve pressure on the French after their cataclysmic casualties—and then withstood the offensives of March and April 1918 when the whole concentrated weight of Imperial Germany, fresh from its victory in Russia, was thrown against the British Army. At the root of that confidence was the fact that the ordinary soldier felt instinctively that, whatever the limitations of his subordinate commanders, Haig embodied the virtues of resolution, calm and consistency, and, above all, of all unshakeable faith, which Britain's amateur army needed to wear down the mighty and supposedly invincible professional force opposed to it and ultimately destroy it. It was not so much on his intellectual gifts— his logical Scottish mind and immense professional expertise—that those who served under him relied as on his character. 'One was conscious in his presence,' the author of this book writes, 'that this singularly handsome man was also a man of rock.'

Writing from first-hand knowledge and observation of his chief gained during the three testing years when the latter was carrying his lonely burden of supreme responsibility, Dr Duncan reveals the faith on which Haig's character was based. He has drawn an unforgettable portrait of him. At the most crucial moment of the German March 1918 offensive, when the Government at home feared that the British Army was broken, this most reticent and taciturn of commanders, replying tersely to an anxious enquiry from the author—then chaplain at his headquarters—surprised him by saying, 'This is what you once read to us from Second Chronicles: "*Be not afraid nor dismayed by reason of this great multitude; for the battle is not yours but God's*".' Alone among Allied leaders in supposing that victory in 1918 was possible, only three months after the issue of his famous Backs to the Wall Order he launched the brilliant and sustained offensive that broke the back of the German Army, winning the greatest military victory in his country's history. 'I always saw that it would come to this,' he said to Dr Duncan two days after the Armistice, 'I never doubted the outcome.' And when the latter tried to congratulate him, he replied, 'Oh! you mustn't congratulate *me*; we have all been in this together all trying in different ways to do our part. It is fellows like him'—pointing to an N.C.O. who had been four years in the front line—'who deserve congratulations.' '*Il était très droit, très sûr, et très gentil,*' was Marshal Foch's summary of this great and much misjudged soldier's character, and this, alas, posthumous book by a distinguished Scottish scholar helps one to realize the justice of that tribute.

PREFACE

I have often been urged to write an account of my association during the First World War with the British Commander-in-Chief in France, Sir Douglas Haig. It is a task which for long I was reluctant to undertake, solely because my relations with him were of too personal a character to be made a matter for public disclosure. Haig was, moreover, a man who studiously avoided publicity, preferring to devote all his powers to the performance of present duty and to leave the future to be the judge of the past; and I should have been untrue to him, not to say untrue to myself, if I had done anything to call attention to a personal association which was to me a unique privilege.

New occasions, however, teach new duties. In recent years many writers have presented to the public a portrait of Haig which is so distorted as to be essentially false. The time would seem to have come for a fresh appraisal of his greatness both as a man and as a commander, and of the services which he rendered to his country in one of the greatest crises in its history.

I have written primarily for the general reader, who may wish to have a picture of Haig from one who saw much of him at close quarters during those years of war. But I trust that the book may also be of service to the serious student, helping perhaps by a statement here, a testimony there, to shed light on the character and achievements of this great soldier.

G. S. DUNCAN

ACKNOWLEDGMENTS

The manuscript of this little book was finished shortly before the death of my husband in April of last year. Had he lived to see it through the press he would certainly have wished to acknowledge his indebtedness to many friends for advice and constructive criticism, and to thank especially Earl Haig of Bemersyde for access to the diaries and other papers of the Field-Marshal; Lord Reith of Stonehaven for continued and helpful interest throughout the writing of the book; and his brother, the Reverend David Duncan, for typing the manuscript and many other instances of his never-failing support. My husband also valued highly the encouragement and practical advice of Mr John Terraine, whose book *Douglas Haig: The Educated Soldier* he greatly admired as doing much to achieve a just revaluation of the Commander-in-Chief. Finally I would like to thank my son, Professor Douglas Duncan of the University of Ghana, for his help in reading proofs.

MURIEL DUNCAN
St. Andrews, April 1966

CONTENTS

The custom that prevailed in old times of passing on to posterity a record of the character and achievements of famous men has not ceased to be observed even in our own day, however indifferent a new generation may be to its past; and when a man of distinction arises, the sinister tendency (found alike in states great and small) to ignore or disparage true worth becomes powerless in his presence—his greatness lifts him high beyond its reach.

TACITUS: *Life of Agricola*, Chapter 1

I : INTRODUCTORY

THE TWO WORLD WARS

Douglas Haig has an assured place in history for the part which he played in the First World War. He crossed to France in August 1914 in command of the First Army Corps, which developed later into the First Army; and in December 1915, with the complete confidence of all who knew him, he succeeded Sir John French as British Commander-in-Chief, and held that post to the end. It thus fell to him, to an extent unmatched by any other commander on the Allied side, to hold without interruption a position of the highest responsibility in the field from the opening of the campaign to its victorious close. He had under his command by far the largest army that Britain had ever sent to war; and while under him the British army, overwhelmed by superior numbers, suffered in March 1918 a very severe defeat, it also accomplished eight months later, in company with the armies of our allies, one of the greatest victories in military history. His name has become associated in the minds of many with the 'blood-baths' of the Somme and Passchendaele; and he has been held responsible for the very heavy losses which the British armies incurred in killed and wounded—though it is to be remembered that, in this most terrible of recent wars, the losses in *all* the belligerent armies were correspondingly severe. On the other hand there are those who maintain that Haig saw with unerring clearness the one way by which Germany's armed might could be broken, and that by his refusal to be diverted from this course and by his ability to force a decision before the end of 1918 he proved himself the real architect of victory. There are good grounds for the belief that with a fuller appreciation of the magnitude of his task and of the special problems with which he had to deal his reputation will mount higher and higher.

The following pages are an appreciation of Haig as I knew him during the three eventful years of his supreme command in France. They are designed to deal above all with his qualities of character, to portray him for those of a new generation who may like to know what sort of man he was. But a man's character cannot be assessed without due regard to historical context; and for a true appreciation of Haig we must try to see him in relation to the critical situations that from time to time confronted him. As we come to recognize better the magnitude of the challenge that the British land forces faced in 1914-18, there stands out in vivid perspective the figure of the British commander who rose to new heights with each succeeding crisis, and brought us in the end to victory. For that reason I have cast part of my narrative in historical and chronological form.

After two world wars it is not surprising that recollections of the second, in its intensity, its horrors, and its triumphs, should have dimmed for a time such interest as would otherwise have continued to be evoked by the story of the first. That time, however, would seem to have passed. As the second war itself recedes into history, we are coming to see how these two wars were in a measure one, separated by an armistice and twenty years of uncertain peace; and, as men in the First War used jokingly to observe, the first seven years of war are always the worst. The two wars had much in common. In both the might of an aggressive Germany was opposed by the allied forces of France, Britain, and the other nations of the British Commonwealth, with the United States of America ranged on their side at a later stage of the campaign. And in both our very survival as a free nation was at stake. But along with these basic resemblances went far-reaching differences; and it may be worth while to recall a few of these. What was the general position on the western front in 1914-18?

(1) France was not completely overrun as she was in the Second War. The first fierce German onrush was stemmed at the Marne before it reached Paris or the Channel ports; and the two opposing armies, with their front lines deeply entrenched in the earth and guarded by barbed wire and a vast concentration of artillery, faced one another for four years in a line that

stretched from the Belgian coast to the Swiss frontier, locked in a grip which (despite ceaseless attempts to weaken it) human ingenuity could find no effective way to break except at a tragic cost in human life.

(2) Russia began in 1914 on the side of France and Britain; but by 1917 she was in the throes of revolution; and Germany was thereafter free to throw practically all her weight against the Allies in the west. For Britain the situation became all the more critical because France was by that time showing unmistakeable signs of war-weariness, and there were doubts whether she would be able to continue the fight.

(3) American intervention in 1917 undoubtedly brought great moral and material support to the Allied cause; but the American army in France was only beginning to exercise its influence as a fighting force when hostilities were near an end.

There was another vital difference between the two wars which cannot easily be expressed in a few words. In 1939 the British people understood the dreadful realities of war; our whole outlook as a nation was clouded by memories of the past and anxieties for the future. To meet the threat of a fresh outbreak we had reconciled ourselves, even in peace-time, to compulsory military service; and when war came we accepted the challenge, unhesitatingly but without elation. It was different in 1914. We had imagined ourselves then to be living in an age of enlightenment; and that a civilized nation like Germany should wantonly provoke a war with her European neighbours came as a shock both to the intelligence and to the conscience. The nation sprang at once to arms; so too did the other peoples of the British Empire. There is no adequate parallel in history, before or since, to the upsurge of stern resolution that the need aroused. The response came from every section of the community. And it was for long an entirely voluntary response; despite the desperate character of the struggle, conscription was not introduced till two years later. Above all, there was an idealistic ardour, a sense of unity, and a comradeship which a later generation has found it hard to understand. Something of the original elation was lost in the awful conditions of the fighting around Ypres and in the

Somme valley; but the idealism remained. This was a war to end war for ever.

All war is tragic. But the 1914-18 war was especially tragic. Among many of our leaders, military and civilian, there was at first no clear appreciation of the magnitude of the task we had undertaken : as against a widespread belief that victory would be ours by Christmas, Kitchener (who had become Secretary of State for War) and Haig were almost alone in insisting that the war might well last three or four years. For such an undertaking we were quite unprepared in regard to manpower, staff or equipment. Our heroic little regular army was soon swallowed up in the fighting of the first twelve months. Its place thereafter was taken by Territorial battalions and by recruits in the new 'Kitchener' armies, men who had left the office or the shop, the university classroom or the factory, and who now found themselves in the battle zone, perhaps even as officers, with no previous experience of war, often with little aptitude for it, and after only a hurried and quite inadequate training. It is no wonder that the best of our nation's manhood was lost in the first two years; and among them were many whose experience and training, had they survived, would have been invaluable in the decisive campaigns of 1917 and 1918.

A further factor in the tragedy of the time was that, having won the war, we lost the peace. The post-war years were years of unrest and disillusionment, in which it became all too easy to see everything out of focus, to depreciate the victory which had been gained by so much effort and sacrifice, and to think of the war primarily in terms of its horrors and its losses. Whereas during the war and in the years that immediately followed Haig was accepted as a national hero both by the men who fought under him and by the nation as a whole, and the devotion paid to his memory at the time of his death had no adequate parallel in recent history, the myth began to take shape, sedulously fostered in certain quarters, that the root-cause of our losses in the war lay in the incompetence and indifference of the British General Headquarters staff, and that Haig in particular showed an almost complete lack of imaginative leadership and an obstinate unwillingness to profit from the lessons of experience. It is a myth which is now at last giving way to a more just appreciation of the facts. We would do well here to turn from it altogether, and by an effort of imagination and understanding to reflect on the immense load of responsibility which, in the testing conditions of that time, had to be borne by all our national leaders, military and civilian alike, and more particularly by Sir Douglas Haig.

With no respite or relaxation Haig was at his post in the field from the opening weeks of the war until its close. It was a 'contemptible little army' that Britain had sent to France in August 1914—it consisted of five divisions; but by the end Haig had under his command the largest army that Britain had ever put into the field, a force of two million men; and within the hampering conditions imposed on him by trench-warfare he grew in stature to meet each fresh demand, and remained (to quote a German verdict) 'master of the field'. In the opening weeks he had rescued his forces from threatened annihilation in the fateful retreat from Mons. Four months later, during the First Battle of Ypres, his First Corps, exhausted by its exertions and with no reserves and only meagre artillery support, faced the full thrust of a fanatical German attack which was designed to force a break-through and to end the war. And the story will live on how, when word reached him that the line was broken, Haig mounted his horse, and with a small escort rode calmly down the Menin Road, his very presence helping to restore confidence. Small wonder that, according to the report of an American who had been in Berlin at that period, the Kaiser gave it as his considered opinion that 'the British First Army Corps under Douglas Haig is the finest in the world'.

In many ways 1915 was a disastrous year. It was the year of Gallipoli—that heroic adventure which, if successful, would have changed the whole course of the war, but which, just because it failed, served to underline the uncertainties and dangers of attempting to achieve a decisive success in a secondary theatre. On the Western front the year was marked by a series of abortive and costly local attempts, by both sides, to accomplish a breakthrough. In the Second Battle of Ypres, which has been described as one of the most murderous battles of the war, the Germans resorted to the use of poison gas, but

B

gained relatively little ground. Protracted French assaults, first in Champagne and then in Artois, accomplished little, with the estimated loss to the attackers of approximately 190,000 men. British failures at Neuve Chapelle and Aubers Ridge were followed in September by the ill-starred Battle of Loos. The obvious misdirection from GHQ in this last battle confirmed the growing belief that a change was necessary in the British Supreme Command; and on December 18th Sir Douglas Haig replaced Sir John French as Commander-in-Chief.

CHAPTER 2

MY ASSOCIATION WITH HAIG: HOW IT BEGAN

My first meeting with Douglas Haig was on the morning of Sunday, January 2, 1916, precisely two weeks after he had taken over the Supreme Command. He appeared at 9.30 that morning at a church service which, as a Chaplain to the Forces, I conducted in the Pas de Calais town of St Omer; and little though I realized it at the time there began then for him and me an association which was to be renewed week by week till the end of the war, to grow in intimacy with the years, and to continue right on till the solemn service in February 1928, when his body was laid to rest within the grounds of Dryburgh Abbey.

Of that first meeting I may say that no one planned it, no one even foresaw it; it just happened. If I recall the occasion in some detail, it is because the whole setting might well have seemed to Haig (if only he had been a different type of man) to be very far from propitious. The scene was a small dingy concert-hall at 116 rue de Dunquerque, which had been allocated to the Presbyterian chaplain as a church; to reach it worshippers entered by a narrow passage into a back court-yard and up an outside iron stair. My senior colleague and I used the hall on week-days as a place of entertainment for the troops; and we did our best at other times to make it not alto-gether unworthy as a place of worship.

As for myself I was a very junior army chaplain; and if I had any qualifications at all for the work, they had been gained in a very different field. I had spent the years 1902-1914 at various universities. Four years at Edinburgh had been followed by three years at Trinity College, Cambridge, my main study at both places being Latin and Greek. My thoughts then turned to theology; and in preparation for the ministry of the Church of Scotland I spent three more years at Edinburgh, a winter at St Andrews, and three summer terms at German universities; and in the summer of 1914, when I was engaged in

post-graduate study of the New Testament at Heidelberg, I was
fortunate in finding my way back to Britain a week before the
outbreak of war. Those were years when I had many opportunities to widen my outlook and develop new contacts. But
they had been by no means easy years. The typical Scottish
student in these times had often to pursue his course the hard
way, dependent financially on what he could gain by scholar-
ships; at Cambridge in 1907 I had undergone a very serious
operation for appendicitis, and was afterwards known among
my friends as a man who had no right to be alive; and for the
next seven years my health remained very uncertain. Perhaps
in this way I learned some lessons other than those to be gained
from books or in academic classrooms.

But from a professional angle I knew next to nothing about
army life when in 1915 I was selected for a commission as a
temporary Chaplain to the Forces. As a minister of the Church
of Scotland I was in the unusual position of never having had
a pastoral charge, and it was in fact only when the offer of a
chaplaincy came to me that I was ordained. For some months
previous to this I had acted as an assistant to the parish
minister of Govan (Glasgow) in his work in that busy indus-
trial centre on the Clyde. But when I was sent to France there
can have been few chaplains who, by previous ministerial training and experience, were less qualified than I was for the work
which a few months later was so unexpectedly to open out
before me.

At that early period of the war (it was different later) a
temporary chaplain was given no special training of any kind
for service with the Forces; he merely received his commission
and was told to get into uniform. And so I crossed to France
in the first week of September 1915, found my way by train
from Boulogne to St Omer, and with half a dozen other new
arrivals appeared the following morning before the Principal
Chaplain, the Rev. J. M. Simms. My companions were all duly
posted, some to base camps, others to units up the line. But
the Principal Chaplain evidently had some difficulty in knowing what to do with me. He asked me to report again the
following morning, when he told me he had decided to retain
me in St Omer, to assist another Presbyterian chaplain (the

Rev. W. J. McConnell of Belfast) in his work with GHQ troops
and other units in the immediate neighbourhood. My work, as
it transpired, was almost entirely in four near-by hospitals, and
among men in GHQ offices, in the signals services, and in a large
motor repair shop. A detachment of the Artists Rifles was also
stationed in the town as an Officers Training Unit. Sir John
French, Commander-in-Chief of the British Forces in France,
had his headquarters in St Omer; but of him and the officers
on the GHQ staff I saw nothing. They moved in an altogether
different realm.

The fact that a few months later Sir John French gave place
to a new Commander-in-Chief had no obvious relevance for
me. Haig took over command on December 19th. He and his
staff duly established themselves in St Omer; and I heard that
on the first Sunday after his appointment he had attended the
Church of England service at what was known as the Military
Church. Then on the second Sunday he came, as I have said, to
the Presbyterian service, one of the first officers of any rank
who had ever found their way to it. As I reflected on the
episode afterwards, I could not help wondering what his
thoughts had been as, accompanied by two ADCs, he took his
seat in that uninviting place of worship. I have often been
asked what my own thoughts were that day. Accustomed as
I had been on previous Sunday mornings to meet with a mere
handful of Tommies, what did it mean to me now to find in
my congregation, sitting within a few feet of me, the
Commander-in-Chief? It is sometimes said of the Scot that he
is not unduly affected by differences of rank and status—'a
man's a man for a' that' is a fundamental element in his home-
spun philosophy. Whether that was true or not of me that
morning I shall not say. It was perhaps due rather to ignorance
and inexperience that I was not more upset than I was. Be
that as it may, I proceeded with the service precisely as I should
have done if Haig had not been present. The day was the first
Sunday of the New Year, appointed by the King to be
observed as a National Day of Prayer; and I should have been
unworthy of my office in the Church if I had failed to treat
the occasion with all due solemnity. It was only when Duff
Cooper's Life of Haig appeared in 1935 that I learned from it

that the Commander-in-Chief had included an account of the
service in his diary. 'One could have heard a pin drop during
the service. So different to the coughing and restlessness which
goes on in church during peace time.'

The question naturally occurred to me: was there any
possibility that he might return the following Sunday? I made
discreet enquiries about this in a certain high quarter, and was
told that it was quite unlikely. I had, however, an easy mind on
that score, for it would then be Mr McConnell's turn to take
the morning service in St Omer while I visited a number of
outlying units. But when Sunday morning came my colleague
was indisposed; with little time for preparation I had to step
in to take his place; and I still recall my trepidation when
once again the Commander-in-Chief appeared and made his
way up that iron stair. Of that service too, as I was to learn
later from Duff Cooper's biography, he wrote an account that
day in his diary. Thereafter he continued to come, not just
occasionally, but Sunday by Sunday; and if on any Sunday he
was not in his place, it was because he was absent from GHQ.
At the end of March 1916 he transferred his headquarters from
St Omer to Montreuil; and I found to my surprise that I was
to be one of those who were to go there.

It was with some apprehension that I viewed the change.
At Montreuil, a charming little self-contained town enclosed
by Vauban's ramparts, there were to be no units of any kind
except those directly associated with GHQ. With no hospital
responsibilities, would I have enough to keep me busy during
the week? A still more anxious question for me was whether
I could expect to have a congregation for worship on Sunday
mornings. On Sunday evenings at GHQ men were generally
free to attend church, and they did attend in good numbers;
but the position which confronted me now (as indeed it
confronted me during all my time in France) was that on
Sunday morning work went on in the offices of GHQ much as
on other mornings; and among those who took time off to
attend church, a very limited number were likely to find their
way at 9.30 a.m. to the Presbyterian service. On my first two
Sundays at Montreuil the congregation was so small that we
read the hymns instead of singing them. Before long, however,

I could count on a congregation of about twenty, including a number of men from the Commander-in-Chief's escort of Lancers under the command of a much loved Scottish officer, Captain George Black, on whose suggestion the men of the escort expressed a readiness to attend worship with their Chief. The Commander-in-Chief himself was invariably present, accompanied perhaps by his Assistant Military Secretary, Lieut. Colonel Alan Fletcher, his Private Secretary, Sir Philip Sassoon, and one or two other members of his personal staff; and a few senior staff officers attended from time to time. Apart from these the congregation consisted almost entirely of private soldiers. There was no formality or parade as the Commander-in-Chief arrived and took his seat in that little wooden hut on the ramparts that now served as our church. As for myself, I was in no sense a staff chaplain. I was simply the Presbyterian chaplain attached to GHQ troops, and as such I remained to the end of the war a chaplain of the lowest grade, unattached to any other unit and with my initial rank as Captain.

Haig's attendance at the Presbyterian service has often been put down to his Scottish descent and his loyalty to the Church of Scotland. This is scarcely correct. True, he had been familiar as a youth with the Presbyterian form of worship; and when the war was over he identified himself definitely with the Church of Scotland, becoming an elder first in St Columba's Church, Pont Street, London, and then in the Scottish parish church near his home at Bemersyde. But denominational loyalties in themselves had little meaning for him. While on occasion he might go to a different place of worship, his practice in the army had been to attend the main military service, which was normally that of the Church of England. He had done so when in 1915 he had been in command of the First Army; he did so too on his arrival at St Omer as Commander-in-Chief; and it is of interest to recall that the very first service he attended there, on Christmas Day 1915, he was so impressed by the sermon preached by his friend, the Right Rev. Bishop Gwynne of Khartoum (who was now Deputy Chaplain-General in France, with responsibility for all Church of England chaplains) that he caused it to be printed for circulation. There is evidence that by this time his religion had become a vital element in his life, and that he had come to view in a more definitely religious light both the issues at stake in the war and the part which he himself was being called to play in it. But he had unfortunately not always received the help he could have wished to get from Sunday morning worship. In a post-war letter to me there occurs a significant sentence: 'I had a hard trial before I came across you at St Omer.' And so it was in something of a pioneering spirit that, on that first Sunday of 1916, having enquired about a Presbyterian service, he went off 'on his own' to worship there.

But what began that Sunday as an experiment became his established practice; and so I was with him to the end. More than once in 1917 and 1918 I asked to be allowed to go elsewhere. I felt strongly that a change would be good for him and good for me, and that I ought to be doing something more strenuous. But Haig hated changes unless necessity demanded; and he was intensely loyal to the men who served under him. And so, when on one occasion during the hard fighting of 1917 I pressed the matter rather urgently, he put his hand on my shoulder and said quite simply: 'If you are of help to me, I hope you will be satisfied.' And to that there could be no reply. He added, however, provided I returned to conduct the Sunday service, I ought to feel free, so far as he was concerned, to go anywhere I wished during the week. This meant that during the summer months I was able to spend a considerable part of my time with units near the fighting line.

CHAPTER 3

THE COMMANDER-IN-CHIEF: A PRELIMINARY PORTRAIT

How did Haig appear to those of us who saw him at close quarters? And what were some of the immediate impressions he left on us?

His appearance was certainly impressive. He was taller than some of his colleagues, e.g. Sir John French, Sir William Robertson and General Foch. On the other hand he lacked the dominating martial presence of Lord Kitchener; his handsome features, always in repose, were at first sight suggestive less of an army commander than of a cultured, gracious and peace-loving country squire. In uniform or out of it he always appeared as 'one of nature's gentlemen'. One was soon conscious, however, in his presence that this singularly handsome man was also a man of rock; his finely chiselled features represented a combination of grace, stability, and strength. Broadshouldered, lithe and erect, his firm stance suggested a readiness to meet four-square whatever winds might blow; and carrying his head high he would give it every now and then a sudden short tilt expressive of conviction and confidence. What rivetted attention most of all as one looked at him was his noble forehead and his luminous blue eyes. Preeminently he was a man of action; but that broad square forehead marked him out also as the calm, concentrated and logical thinker, and the look in his eye suggested unplumbed depths of feeling and understanding. As he spoke to you, you realized instinctively that here was a man of transparent honesty, who for all his reserve had a warm heart, who in all his judgments was accustomed to look below the surface, and who brought to all his severely practical tasks a strong measure of far-sightedness and idealism.

Major Neville Lytton, in his book *The Press and the General Staff* (1920), gives a picture of Haig which deserves to be quoted. After being wounded in the Somme battle Major Lytton had been selected to be head of a foreign press mission attached to the British army; and he tells as follows of a meeting he had with the Commander-in-Chief.

'I don't think,' he writes, 'I shall ever forget the impression that the Chief made upon me; it was the first time that I had ever seen him, and I fell immediately under the spell of his personal magnetism. . . . I confess that I am one of those who will do anything for one sort of man and nothing for another; well, with Haig I felt immediately such a longing to gain a word of praise from him that I would have liked him to ask me to do some impossible exploit that I might prove my devotion to him. . . . Haig's qualities are much more moral than intellectual; what intellectual qualities he has have been used almost entirely within his own profession, but he exhales such an atmosphere of honour, virtue, courage, and sympathy that one feels uplifted like as when one enters the Cathedral of Beauvais for the first time. Surely it is this sense of trust that has made Haig come out on top in spite of a terrible rough passage; not one of his subordinates has ever suspected that he could act from any motive of self interest. It is largely the great character of the Commander-in-Chief that brought about the astounding result that Great Britain had the finest field army of the world in the autumn of 1918.'

From my own experience I would endorse every word of that verdict. I never met Haig at any time without feeling immensely better for being in his presence. Every time I came face to face with him I was struck by the genial, kindly look in his eye. He always looked you straight in the face; and, having as a rule little to say, he would make up for this by giving you a welcoming handshake. He was, it is true, undemonstrative, and always remained in a measure aloof. This was due in part, no doubt, to a streak of shyness in his nature. But there was also, I suggest, a deeper reason; being true to himself, he felt no need to 'play up' to others, he had no desire to try to create a good impression.

His reserve, combined with his lack of ease in conversation, was, of course, a serious handicap in dealing with certain types of men, or with men in the mass. He had, for example, none of the dazzling volubility with which his army colleague, Sir Henry Wilson, readily gained the ear, and the confidence, of Mr

Lloyd George. So too he was without those popular gifts which have enabled some military leaders, especially those with a field command, to establish close personal relations with the troops. Yet there was something in Haig which undoubtedly inspired confidence. Those who worked in closest association with him were enthusiastic about him; among those who knew him only at a distance, and indeed throughout the army as a whole, there was a recognition that he was a man to be trusted to the full. What won most men who approached him without prejudice was his disarming sincerity—'a deep, great, genuine sincerity' which, as Carlyle writes in his *Lectures on Heroes*, 'is the first characteristic of all men in any way heroic'. Haig always rang true.

Haig's reserve was no doubt liable to be misunderstood. But there is no justification for the false presentation of it which has gained currency in recent years. He has been described as cold, self-centred, lacking in human understanding and sympathy. I have even seen it stated that few men in public service have been more plagued by an unattractive personality. This is gross caricature, the product of ignorance and prejudice. As in the post-war years he went from city to city, from land to land, caring for the needs of ex-servicemen and their dependants, the whole world came to know Haig as one of the most sensitive and kind hearted of men. And beneath the surface there was the same depth of feeling, kept under strict control, in the years of war. Here, for example, is a tribute which is worth recalling, coming as it does from that discerning judge of men and affairs, Mr F. S. Oliver, author of *Ordeal by Battle*, who first came to know Haig when he visited him in France in 1917. In his posthumously-published volume of letters, *The Anvil of War*,[1] Oliver has this to say of Haig: 'I think he is about the least self-seeking creature I ever ran across. And certainly, as I see him, he is one of the warmest hearted and most sympathetic. I've met and known fairly intimately most of the British bigwigs who are engaged in this war. As far as personal affection goes I put D. H. first.'

Haig took men as he found them; his honesty of approach,

[1] p. 306.

together with his readiness to believe the best, went far to create an atmosphere of understanding. I think, for example, of his reactions to Lord Northcliffe, whom Mr Lloyd George had asked him to entertain in the course of the Somme battle. Haig had anticipated the visit of the press lord with some apprehension; but shortly afterwards we find him writing in his diary: 'I quite like the man; he has the courage of his opinions and thinks only of doing his utmost to win the war.' And this is typical of Haig's general attitude in human relationships. I have watched him talking to a private soldier engaged in some menial duty; and though he might have little to say he spoke as man to man, and succeeded in conveying the impression that he cared.

An interesting reminiscence comes to me from a former member of the Honourable Artillery Company, that fine body of men who so often provided the guards for GHQ. On one occasion an inexperienced HAC recruit, engaged on some duty near the entrance to the Commander-in-Chief's chateau, had turned aside for a moment from his task to glance at a map in the hall. To his dismay the Commander-in-Chief chanced to appear on the scene. Haig attached importance to the observance of discipline; but his deeper nature revealed itself when he said to the offender: 'You are interested in maps, I see. When you are not strictly on duty, come inside and let me show you some of my more recent maps.' The invitation was so obviously sincere that the proud if somewhat frightened private felt he must not fail to accept it.

No small part of Haig's reserve came from his absorption in the work he was given to do. In a sense he was never off duty. There was something in this which was not without meaning for the men in the trenches. They may never have seen their Commander-in-Chief; he was perhaps little more than a name to them. But his reputation had come down from the 'old contemptibles'; they valued him all the more because he never sought to strike a pose. Mr Ernest Raymond,[1] author of *Tell England*, who served as an officer at Passchendaele, tells how the troops reacted to this side of Haig's personality. 'Some-

[1] *British Legion Magazine*, March 1928, p. 266.

times I fancy that the name of Haig will be one of the most picturesque names in our history, by reason of his very lack of picturesqueness. . . . The huge temporary army of 1914-1918 disliked picturesqueness and melodrama; they were simple, direct people, a quiet obstinate crowd, whose chief merit, I hold, was nothing more than their obstinacy; and they knew, without knowing that they knew, that they were commanded by a man after their own heart.'

Perhaps the most notable element in Haig's character (it was indeed part of his natural genuineness and sincerity) was his resolute sense of purpose. One could not be long in his presence without being conscious of this. It had indeed been a marked feature of his character ever since he entered the army. He had been a late developer. Leaving school at Clifton with no definite ideas about a career, he had travelled across America as far as California, and then spent three years of glorious irresponsibility at Brasenose College, Oxford. Finally, at the age of twenty-two, he entered Sandhurst, a few years older than most of his fellow cadets. He had come from a family with no great military traditions; but as a Sandhurst cadet and later as a commissioned officer he became a marked man, both by his ability and by the thoroughness with which he set about preparing himself for such work as might lie before him. He seemed to see ahead a day of reckoning in which the British Army would be severely tested; and no kind of training, study, or discipline was regarded as too exacting if it would make him, and the troops entrusted to him, better equipped to meet the challenge.

Haig was still a captain when at the age of thirty-eight he left the Staff College, senior in years but junior in rank to many of his contemporaries who had gained promotion through active service; but he had already been hailed by a discerning judge at the College as a future Commander-in-Chief. Immediately thereafter he showed his worth, first in the Sudan, and then in the South African war, where he was Chief Staff Officer under Sir John French in the Cavalry Division. Impressed by his abilities Kitchener, now Commander-in-Chief in India, insisted on having Haig, though still a colonel, appointed to his staff as Inspector-General of Cavalry; and within a year he was major-general.

The success of his work in India, where he brought the
cavalry, both British and Indian, to an altogether new level of
proficiency, marked him out for still more responsible duties.
Faced with the growing threat of a European war the Government recognized that our whole army system called for radical
reorganization; and Haldane, who as War Minister brought all
his powers of disciplined thinking to this stupendous task,
decided after careful enquiry that Haig was the man best fitted
to be his military adviser. For three years these two men, so
dissimilar in their general interests and training, worked
together in complete harmony, and, despite serious opposition,
carried through a detailed scheme of reform which included
(a) preparations for an Expeditionary Force to serve, if need
be, overseas; (b) the creation of the Territorial Army as the
second line of defence; (c) arrangements by which the
Dominions were to organize and equip their troops for service
in a unified imperial army; and (d) the inauguration of an
Imperial General Staff.

Haig had good cause too to look back with pride on this work
of reorganization, which ensured that in 1914 the army was
ready, within the limits of size and equipment, for the fateful
test which lay ahead of it. He often spoke to me of this. He
liked to point out how much it had meant to the army that it
now had a clearly defined purpose; it existed not merely to
defend our shores from invasion, but to take part if need be in
war overseas. With this purpose expressed in the very constitu-
tion of the army the soldier now had an aim which he could
keep steadily in front of him, and work to achieve. In our
conversations, however, Haig never dwelt on the part which he
himself had had in this momentous reorganization. Always the
talk centred on what Haldane had done; and for Haldane no
praise was too high. He had been 'the saviour of our country',
'Britain's greatest War Minister'.

Released in 1909 from the War Office Haig went first to
India, as Chief of the General Staff there, then to the important
command at Aldershot, devoting himself in both places to
elaborate schemes of army training for the war which now
seemed to be not far distant. When finally war did come, no one
in the British Army was so well prepared for it as Douglas Haig.

CHAPTER 4

LIFE AT GENERAL HEADQUARTERS

With the growth of the British Army in France and the near approach of the Battle of the Somme it was a strategic move when, in the spring of 1916, Haig transferred General Head-quarters from St Omer to the little town of Montreuil, some twenty miles south of Boulogne near the main road to Amiens and Paris; and there GHQ remained till the end of the war. Its fuller name, Montreuil-sur-mer (it is the M.-sur-M. of Victor Hugo's *Les Misérables*) is a reminder that it was at one time accessible from the sea; and merchants came by boat to dispose of their wares in what was then a prosperous city. But for many centuries now the river Canche, once tidal, has been a muddy stream; to reach the sea at Etaples and Le Touquet is a journey of some ten miles; the peace-time population of Montreuil has shrunk to under 3,000; and much of its earlier glory has departed.

Nature and history have however combined to give the old-world town a distinction all its own. Perched proudly on a hill-top and serenely secure behind its ancient walls and ramparts, Montreuil is indeed an abode of peace; and artists from many lands rejoice in the quaintness of its old tiled roofs and cobbled streets, and in the varied beauty of the landscape. But the natural strength of its position has exposed it throughout the centuries to the stresses and destructions of war; and as a sentinel town in that north-eastern corner of France it played no small part in English military history. The victorious English yeomen on their way from Crécy in 1346 passed under its walls without molesting it; their successors two centuries later made an unsuccessful attempt to besiege it; and in 1804 it became Marshal Ney's headquarters in preparation for Napoleon's projected invasion of England.

As a *locus* for our General Headquarters Montreuil was at once central, isolated, and self-contained. Sentries guarded the barriers at its two approach roads, and no one entered without a pass. Its Ecole Militaire, built as a hollow square with a narrow entrance gateway (where again one might be called on to show a pass) provided admirable accommodation for the main GHQ offices, including those of the General Staff (Operations and Intelligence), the Adjutant-General, the Quartermaster-General, the Artillery Adviser and the Engineer-in-Chief. Other offices were scattered throughout the little town or in near-by villages; and a number of departments (forming GHQ 2nd Echelon) were stationed at the town of Hesdin, sixteen miles away. When early in 1917 a new Transportation Directorate was set up, to deal *inter alia* with railways, roads, docks, and inland waterways, a new village (a well-planned series of huts and corridors) was established a few miles from Montreuil. Its Director-General, Sir Eric Geddes, and the heads of all the main departments were drawn directly from civil life, and given high military rank—an arrangement which aroused opposition and amusement in certain quarters but had Haig's vigorous backing.

In the reach and range of its operations GHQ as it existed at Montreuil represented something new in British military history. It was the Whitehall of the army in the field—excluding outlying departments its population (officers, clerks, guards, orderlies) might be put at 5,000. And like Whitehall it was a collection of units rather than a unity, too large and amorphous to have a well-ordered corporate life. Responsible officers tended to be on duty from nine in the morning till ten or eleven at night, often with no break except for meals and a short sharp walk around the ramparts. If further time was wanted for exercise it could be gained by curtailing the lunch interval. One officer with whom I enjoyed an hour's walk once a week while he was at Montreuil was A. D. Lindsay (afterwards the Master of Balliol); and on these occasions he and I had a snack lunch for which we allowed ourselves fifteen minutes. There was certainly no idleness at GHQ; and it is not surprising that some men broke under the strain.

Another unfortunate aspect of our life was that members of one unit had few opportunities to meet members of another, though this was happily rectified by the establishment in 1917 of an excellent Officers' Club. I may add that we led a monastic life. In the first year of the war there had been critical comment in Parliament about lady visitors to GHQ; but under Haig's command no lady was given entrée, until in 1917 a few members of the Queen Mary's Auxiliary Army Corps (QMAAC) were introduced as waitresses at our Officers' Club, and later permission was given for three Scottish ladies to assist me in the running of our Soldiers' Club, the Scottish Churches' Hut.

I am drawn here to refer to the Scottish Churches' Hut, for it provided a centre for much of my work at GHQ. When we first moved to Montreuil the Camp Commandant assigned me as my church a disused army hut which he had found on the ramparts near the way to the ancient Citadel. 'I bought it for fifty francs,' he added, 'the cheapest church I've ever known.' And it was to this unpretentious structure, picturesquely guarded by trees like a country church at home, that the Commander-in-Chief came for worship every Sunday morning when he was at Montreuil. But the hut soon acquired an additional reputation. It became, under the auspices of the Scottish Churches, a Soldiers' Clubroom, the only one of its kind within the walls; and it was soon the recognized social centre for GHQ troops.

By a happy chance there had come to live in Montreuil some years before this a well-known member of the Glasgow Art School, Mr R. Macaulay Stevenson—a weird genius, wholly unconventional and unrestrained, and with a heart of gold; and not merely was 'Macaulay' proud to serve on Sunday mornings as my 'beadle' (church officer), but he and Mrs Stevenson presided for two years over the day-by-day running of the hut, bringing their artistic and other gifts to bear on making it without question one of the most attractive soldiers' clubs in that northern part of France. Among the regular visitors to the hut was Corporal Howard Spring, then serving as a clerk in the Intelligence Department; I little realized then in my frequent talks with him the reputation he was later to win for himself in the journalistic and literary world. And in one of his books, *In the Meantime*, where he devotes some pages to his life at GHQ, he dwells in memory with warm appreciation on what he owed to the Scottish Churches' Hut.

For seven days in the week, except for the brief intervals on

C

Sunday when it was reserved for worship, the Hut was given over entirely to the social needs of the troops. But by 9.30 on Sunday morning a transformation had taken place; tables and refreshments had been removed; at the far end stood the Communion Table, with a covering of khaki cloth on which was sewn a simple white cross; and the atmosphere was one of reverence and peace. I recall on another page[1] the sense of peace and resolution which Haig confessed to be his so soon as he crossed the threshold and passed into the church. And when, in March 1919, his command in France was coming to an end and the days of the Hut were numbered, he sent the following farewell message: 'If, as I confidently believe, there will be few among the men stationed at Montreuil who will not carry away with them as one of their most pleasant recollections the memory of the Scottish Churches' Hut on the Ramparts, I feel that I too owe it a debt of gratitude. . . . I know that the influence for good exercised by it during the past three years has been immense. I trust that its spirit will live among all who now leave it, and its influence outlast the Ramparts of Montreuil.'

Haig had his personal headquarters at Beaurepaire, an undistinguished chateau in its own grounds some two and a half miles out of town on the highroad towards Hesdin. With him there he had his Assistant Military Secretary, Col. Alan Fletcher, his Private Secretary, Sir Philip Sassoon, and the Chief of the General Staff, Lieut. General Sir Launcelot Kiggell. Living near at hand were other members of his personal staff, including four or five ADC's, who when they were not in personal attendance on the Commander-in-Chief were free to be sent off on special missions and to bring back reports. Some of these were senior officers incapacitated by wounds or ill-health; among younger men who served for a brief period so as to gain experience were General Sir William Robertson's son Brian (now Lord Robertson) and Desmond Morton, who was closely associated with Mr Winston Churchill in World War II. It is of interest to note that Haig had his own Medical

[1] p. 127.

Officer, a jolly Irishman, Col. 'Micky' Ryan. Ryan was indeed
a personality. A regular officer in the Royal Army Medical
Corps, he had crossed to France with Haig in 1914; and his
services during the Retreat from Mons and in the First Battle
of Ypres had been so meritorious that Haig in 1915 got him
attached to the HQ staff of the First Army. General Charteris
tells in his diary[1] an interesting story about this episode. 'Ryan
comes to us today as Medical Officer. We find we cannot do
without him. The immediate cause was General Hobbs developing appendicitis, whereupon D. H.—who believes that the
medical profession comprises only Ryan and a few learners—
telegraphed for Ryan, and now will not let him go. I am very
glad, for he is not only the best of companions, but has the
quite invaluable faculty of making every patient fully convinced that there is nothing whatever the matter with him.
He is also developing a tendency to bully D. H.—which
is very salutary.'

Later Ryan came with Haig to GHQ and remained with him
to the end. I do not know how his duties were defined. But he
was an invaluable member of our GHQ fraternity, welcome
in every company, and always ready to undertake some special
commission. His greatest service of course lay in the influence
he exercised over the Chief. Haig was no valetudinarian—the
fact that he carried on during the whole period of the war with
scarcely a day's illness is sufficient evidence on that score. But
it meant much to him to have close at hand a man like Ryan
who could keep a watchful eye on him, and whom he had
come to trust implicitly, not merely for his professional gifts,
but above all for his robust and cheerful attitude to life and
his unfailing commonsense. In his genial way Ryan continued
to 'bully' Haig as no one else would have dared to do, laying
down the law when occasion arose with regard to food, hours
of sleep etc.; and the Chief would meekly reply: 'All right;
I'll do as you tell me.'

Haig was careful to keep fit by regular exercise, more
especially by riding. On his afternoon rides he always had with
him an ADC and some members of his personal escort (from the

[1] At G.H.Q., p. 72.

17th Lancers); and after an hour's hard ride he would dismount and complete the last few miles of the way home on foot. When other duties imposed on him a long car journey, he would often send horse and escort ahead to meet him at some prearranged point. He never allowed the day's events to 'get him down'. It was as if he said to himself: 'Things may be bad today; but they may be worse tomorrow and in the weeks and months ahead.' He had disciplined himself to take each day as it came.

Each day's programme was carefully mapped out. He was at his desk by 9 a.m.; and normally a great part of the morning was given over to interviews with heads of departments. Among those who reported daily were General John ('Tavish') Davidson, of the Operations Office, and General John Charteris, Chief of Intelligence, two comparatively junior men for such posts. They had been with Haig in the First Army, and he placed great reliance on their capabilities; though as the war went on they came under a good deal of criticism both in Government circles and in the field. Certain days in the week were devoted to conferences with Army Commanders and to visiting forward areas. But as Commander-in-Chief Haig's responsibilities were not limited to operations against the enemy; they extended to the whole field of army organization. In a war on this scale, for which Britain at the outset was so ill prepared, constant attention had to be given to the supply, training and equipment of men for the line, to the provision of munitions of the right types and in adequate quantities, and to the development of auxiliary services (e.g. the Air Force, Tanks, and Transportation). The Commander-in-Chief was the representative and advocate of the army in the field vis-à-vis the home Government on the one hand and the responsible authorities, military and political, of the Allies on the other; and no small part of Haig's greatness as a commander lay in the foresight and pertinacity with which he carried out his tasks in this difficult role.

He was seldom seen in Montreuil except at church on Sunday mornings. After the service he frequently walked round to the Ecole Militaire to pay a personal call at some of the offices. He was generally credited with having a sound grasp of what was going on in the various departments, and with being quietly but sincerely appreciative of good work well done. But he did not obtrude where his presence was not needed. Haig was one of those commanders whose best work was done unobtrusively. I never heard the least complaint that, though he lived so near at hand, he appeared so little among us in Montreuil. Rather there was a real excitement when he did appear. Everyone, including the most junior officer as well as clerks and orderlies, had a deep-seated and spontaneous regard for Haig on personal grounds. We all seemed to realize too the immense responsibilities which he carried; and felt pride in seeing that he carried them with such composure.

II : HISTORICAL

CHAPTER 5

1916 (i): THE FIRST SIX MONTHS

1916, Haig's first year as British Commander-in-Chief, was to see some of the fiercest fighting of the war. After the fruitless battles of the previous year both sides were determined to go all out to force a decision. They little realized as the year opened that there still lay ahead three years of agony and bloodshed.

The German commander, Falkenhayn, got his blow in first. Instead of attempting a big frontal attack on the Allied line he daringly began on February 21st a massive assault on the French frontier fortress of Verdun, hoping (as he said) to compel the French to bleed to death in its defence. He achieved his aim in one respect; for in the ensuing five months the Verdun 'Hell' (as both sides called it) involved France in the loss of approximately 300,000 men, a shattering total to be added to the very heavy losses of the previous years. German losses in the battle were scarcely less severe; and Verdun still remained in French hands.

Allied plans took a different form. Haig's personal preference would have been for an attack in Western Flanders; but in accordance with the War Office directive that the British Army should always act in closest cooperation with the French, he had acquiesced in General Joffre's decision that a sustained Franco-British attack should attempt a break-through in the valley of the Somme. Time, however, would be required for preparation, and not least for the training of Britain's 'citizen' troops. 'I have not got an army in France really,' Haig wrote at the time, 'but a collection of divisions untrained for the Field.' As the threat to Verdun became more serious, and casualties continued to mount, France not surprisingly looked to Britain both to help her in her present emergency and to undertake the main burden of the Somme offensive. Haig was thus confronted with a series of conflicting demands. With characteristic cordiality he acceded to Joffre's request to relieve the French 10th Army; and at the same time he hurried on his plans, despite certain serious shortages in artillery, for the big attack to open if possible in the end of June. On a realistic view of the situation the prospects of a decisive Allied victory in 1916 were now far from bright. But a factor which greatly influenced Haig's calculations at this time (as it did increasingly in the years that followed) was the question how long France would be able to hold out; he had been assured from many sources, including General Joffre and President Poincaré, that one more offensive was all that France would be fit for. He determined accordingly to attack with all his strength in the hope of achieving a success which would both relieve and encourage the French and bring nearer the day of victory.

I turn from those momentous happenings to tell of my personal relations at this time with the Commander-in-Chief.

For some months after my first meeting with him in St Omer I saw nothing of him apart from his regular appearance at church. The church service was apparently a matter of real significance for him, if we may judge from the references to it which he now began to make Sunday by Sunday in his diary. But while we were at St Omer my senior colleague and I had no clue as to why he chose now to come always to the Presbyterian service; and we would sometimes ask ourselves wistfully: 'is he likely to continue doing so?' The fact remained, however, that he did continue to come, and nothing could have been more gracious than his handshake, with a few kindly words of appreciation, at the close of each service.

A significant development came at the end of March when I was posted (and I was told this was on Haig's instructions) to the new GHQ at Montreuil. My official position was still unchanged; I was the Presbyterian chaplain to GHQ troops; but I was now generally accepted (unofficially) as the Commander-in-Chief's chaplain. Sometimes I wondered whether I ought to pay a call at his chateau; but I hesitated to do so with-

out being invited. We had been at Montreuil for a month or
so when, in the course of a friendly talk before the church
service, he told me that he was looking forward to a visit from
the Lord Provost of Edinburgh and the Lord Provost of Glasgow; and then he added: 'perhaps you would care to come
to dinner to meet them.'

I felt I had indeed reached a new stage in my personal relations with the Commander-in-Chief when that evening came,
and I sat down for the first time as a guest at his table.

It was often said by those who visited Haig at his chateau
that, though he was not an effusive host and sometimes had
little to say to his guests, he always welcomed them with an
unaffected courtesy which greatly delighted them and put them
at their ease. This was certainly my experience that evening.
When I arrived he came personally to greet me, and invited
me to join him in his working-room, where everything, as I
noted, was in perfect order; then, after a break in our small
talk, he picked up from his desk some interesting trench photographs which had arrived for his inspection and explained to
me what was significant about them. At dinner, where we
were joined by Generals Kiggell and Butler of the General Staff
and General Charteris, Chief of Intelligence, he was content
to take a minor part in the conversation; but he joined with
relish in the talk of the two Lord Provosts, both men of strong
individuality, who had many amusing stories to tell, and who
left him in no doubt that the Scottish people were whole-
hearted in their support of the war-effort, and would see it
through to victory.

Dinner over, we adjourned to another room, where we dis-
cussed a variety of topics including the food situation in Britain,
France and Germany, and the troublesome question of hours
and wages in the factories and workshops at home; and after a
time our host asked leave to bid us good-bye, as he had other
matters to attend to. There was nothing unusual in this; he
generally, as I came to know, did some work after dinner, perhaps having a talk with CGS, General Kiggell, or writing up an
account of the day's proceedings. That evening he had a good
deal on his mind, as M. Clemenceau, the veteran French states-

man (nicknamed the Tiger) had requested an interview which was to take place the following day; and Haig knew he would have to walk warily. But of all this no hint was given to us that evening; Haig never allowed his official anxieties to cloud the genial welcome due to his guests.

An interesting story lies behind this meeting with M. Clemenceau. It helps us to see how Haig could handle a delicate situation with tact and perfect frankness; it reveals too the place of honour and trust which he had won for himself in his relations with the French. Clemenceau's original proposal, received two days before, had been to call personally on Haig at Montreuil. But, knowing that the French statesman was likely to raise doubts about Joffre and the coming offensive, Haig found a pretext for delaying the meeting until he could have an opportunity to tell Joffre of the proposed interview, and to review the whole situation with him. So he set off at once to see Joffre at Chantilly. As he tells us in his diary: 'The old man was quite delighted to see me; and when I made some ordinary remark about the day clearing up, he said: *Il fait toujours beau temps quand vous veniez me voir.*' The way was thus cleared for the talk that was to follow with Clemenceau. Haig had courteously agreed to come to meet him in a forward area, and in an intimate talk which lasted over an hour he succeeded in putting Clemenceau's mind at rest on some matters that were giving him great concern. Haig writes of the talk: 'I found him most interesting, and we parted quite friends; for, as the proverb says, Friends are discovered, not made.' Between these two men, so unlike one another in many ways, there was begun that day not merely a friendship but a deep-seated confidence and trust which was to grow with the years, and was to prove in the end a factor of incalculable influence in the maintenance of Allied unity.

As summer drew on it became common knowledge that a big attack was to be launched soon. And on the last Sunday in June the Chief said to me at the close of the service at Montreuil: 'Well, we'll see you at Beauquesnes next Sunday.' (Beauquesnes, an agricultural village between Albert and Doullens, had been selected as an Advanced GHQ.) This took

me by surprise; and I asked in some bewilderment: 'But
mustn't I remain at my post here?' 'Oh! I'll see to that,' he said
with a disarming smile. And so I was duly posted to Advanced
GHQ, and the Principal Chaplain made temporary arrangements for the carrying on of my work at Montreuil.

Life at Beauquesne promised to be very different from that
at Montreuil. Only a very limited number of officers and men
had been moved to this advanced station; and I was to be the
only chaplain 'in residence'. It was not immediately obvious
how I was to spend my time. I could not doubt, however, that
with the opening of the offensive work in abundance would
soon be found for me.

At Beauquesne I was at least to have some interesting
personal contacts. On one of my first afternoons I was walking
along the mean village street when I was hailed from the
opposite side by General Charteris, Chief of the Intelligence
Department. He had two friends with him whom he said he
wished me to meet. And turning to the first he introduced Lord
Crawford. Face to face with me I saw the Earl of Crawford and
Balcarres, the Premier Earl of Scotland, undoubtedly one of the
most learned members of the British Peerage; but being at this
particular moment a corporal in the Royal Army Medical Corps,
Lord Crawford immediately clapped his hands to his side and
stood rigidly to attention, so that I was at a loss how to proceed.
Then came the second introduction: 'This is John Buchan,'
and Buchan, being dressed as a civilian, readily shook hands,
and conversation was easy.

To this little story there is a background which is worth
recalling. All three had come from seeing Haig. Hearing how
Lord Crawford was acting as an RAMC stretcher-bearer,
Charteris had 'dug him out' to offer him a post as Lieutenant
in the Intelligence Corps. But in that very week the Prime
Minister had decided to make him Minister of Agriculture with
a seat in the Cabinet. Lord Crawford's own hope, I understand,
had been that this visit to GHQ would settle the issue, and that
the Commander-in-Chief would call for his retention in France.
But clearly the last word was with the Government; and home
he had to go. From 1916 to 1918 he was Lord Privy Seal.

John Buchan's presence in France had arisen out of a grow-

ing demand at home that much more should be done to boost
the British war-effort both among our own people and in
neutral countries. He had arrived that morning from England, with the authority of the Foreign Office, and was to
remain in France as a commissioned officer for specialized duties
in connection with 'press' and 'propaganda'. It was work
that created many problems, for the Foreign Office and the
War Office were not always of one mind as to what sort of
information should be revealed to the public and what with-
held. Some three months later Buchan was asked by GHQ
to undertake another duty. The Intelligence Office had been
given responsibility for the preparation of the daily war-
communiqués; and I recall hearing from General Charteris
that, being pressed to make these rather more vivid and excit-
ing, he immediately said to himself: 'What more suitable man
can I get for this than John Buchan, if the Foreign Office will
release him?' As for myself, that first casual encounter was the
beginning of many happy associations with Buchan during his
time with the army, to be renewed in later years when he
became MP for the Scottish Universities, and when in 1933 and
1934 he represented the Sovereign as Lord High Commissioner
to the General Assembly of the Church of Scotland.

On June 29th, two nights before the Somme battle opened,
I dined with the Commander-in-Chief at his new quarters, the
Chateau de Valvion, on the outskirts of Beauquesne village.
We were a small party; and as we sat at dinner there came
from an adjoining room sounds of a piano—a little girl was
playing her five-finger exercises. I thought by contrast of the
'sound of revelry' that preceded Waterloo. Here was a
Commander-in-Chief who, rather than cause undue inconven-
ience to the legitimate owner of the house, was willing that
she and her little daughter should continue in possession of a
few rooms.

CHAPTER 6

1916 (ii): THE BATTLE OF THE SOMME

After a week's fierce bombardment the battle of the Somme opened on July 1st. The British attacked on an eighteen mile front north of the river, with strong and effective French co-operation to the south. The opening day was in many ways disastrous; there was no break-through, and British losses in killed and wounded amounted to 57,000. Much has been written and will continue to be written on the Somme battle, and in particular on that opening attack. I content myself here with telling in a few sentences how Haig, as I believe, viewed his problem, and how in the end his plans were foiled.

The first necessity was of course to open a way, by intense artillery fire, through the enemy's front line defences. What thereafter was all-important (as the failure at Loos had clearly shown) was that without loss of time the infantry should pour through the gap in one steady continuous advance. Was Haig here expecting too much of a citizen-army—an army moreover in which so many senior officers in highly responsible commands were without adequate training and experience? He was certainly misled by over confident and grossly inaccurate reports from some of his Corps Commanders; and troops advancing to the attack found that the wire opposite them was in some places imperfectly cut. What was even more serious, the enemy, with characteristic ingenuity and thoroughness, had developed a system of deep concrete dug-outs which were untouched by artillery fire; and when the barrage lifted they emerged with machine guns to mow down our men as they advanced in wave after wave. And in 'the fog of war' there was no means by which those behind could keep contact with the situation as it developed ahead. Walkie-talkie sets were not available then as they were in World War II.

On that first black day of the Somme battle I went, on the Chief's suggestion, with Major Thompson, one of his ADC's, on a round of visits to headquarters and hospitals behind the battle front. The part of the line allocated to us lay south of Albert, where, as it happened, casualties were less severe than they were elsewhere; and for my part I returned that evening with very little realization of the magnitude of our losses on that opening day. A week later it was arranged that, in addition to my duties at Advanced GHQ, I should act as chaplain to two Casualty Clearing Stations at Puchevillers, some three miles away; and for the next four months I lived there under canvas. During that time I found this double duty extraordinarily challenging in the demands it made for sympathy and understanding. Engaged as I was from Monday morning till Saturday night with the wounded, the dying and the dead, and in writing letters to their relatives at home, it was with a new experience of the tragedy of war that I returned on Sunday morning to Beauquesne, to meet with the Commander-in-Chief and those others who gathered with him there for Christian worship.

I was going about my hospital duties one day when I received intimation from the Chief that M. Poincaré, the President of the French Republic, was to call on him on Sunday morning at 9.30. As I read the opening sentence of the letter my first thought was how characteristically courteous it was of him to let me know that he would not be free to attend the church service. But as I read on I found that his purpose in writing was to ask whether I could change the hour of service that day till twelve o'clock, so that he might come after the President had departed. I need not say that this letter gave me a new insight into his attitude to Christian worship. I was now familiar with his regular attendance at church, which I could see was by no means a matter of convention. But that on an occasion such as this, when another important duty made it impossible for him to attend, he should be so unwilling to miss the service that he would rearrange his day's programme and come at a later hour, there was something there for which I had not been prepared. It helped me to see that the half hour of worship was not merely something which he valued; it was apparently something that must not be missed. He had never previously opened his mind to me on such matters; but as I pondered that letter I seemed to see that for him the war was more than the clash of armed nations, it was a conflict in which

the divine over-ruling was at work; and military might, for all
its importance, was in the long-run of little avail apart from
the resources of the spirit.

With swaying fortunes and only limited successes the Somme
battle dragged on till November. On Sunday, November 12th,
the Chief was at church as usual; and I could not help noting
his resolute look as he shook hands before going out. There
was something in his manner which seemed to suggest that he
was prepared in spirit for the tasks to which he was now to
address himself. He went straight from church (so I learned
later) to Fifth Army Headquarters at Toutencourt, to confer
with General Gough on the projected attack on the Beaucourt-
Beaumont Hamel ridge. Something of what would be involved
in the attack was set down by him later that day in his diary;
it may be read in his published *Private Papers*. A success would
strengthen the whole British position before the winter set in.
'But the necessity for a success must not blind our eyes to the
difficulties of ground and weather. Nothing is so costly as a
failure. But I am ready to run reasonable risks. I then dis-
cussed with Gough what these risks were, and why he thought
our chances of success were good. Finally, I decided that the
5th Army should attack tomorrow.' I recall one GHQ officer
saying to me afterwards that the attack on Beaumont Hamel,
strictly limited though it was in scope, was one of the big
decisions of the war to date. The attack opened in dense fog
at 5.45 on the morning of Monday, November 13th, and was
completely successful; and this was also at a cost significantly
less than was involved in the earlier failure to capture the same
place in July, a sign how much the army had gained in fighting experience since then. But in the awful conditions of rain
and mud there could be no question of pursuing the success
further.

With this conclusion to the large scale offensive on the
Somme the Commander-in-Chief and his staff returned that
week to Montreuil.

The Somme did for the British people what Verdun, and the
fierce fighting of the previous years, had done for France; it

stabbed our country wide awake to the grim realities of modern war. There was nothing in Britain's earlier history to parallel this experience—a battle continuing for months under the most gruelling conditions, and fought, not as in the old wars with a comparatively small body of mercenary troops, but with an army recruited mainly by voluntary enlistment. This was mass-slaughter. It was so, it must be remembered, not for Britain alone, but for all the nations engaged; it has been estimated that the German army lost more men on the Somme than the British and French together. For Britain the loss was not to be reckoned merely in numbers; it meant the sacrifice of the pick of her manhood. Casualties were especially severe among junior officers. I think of many whom I had come to know intimately among the Artists Rifles at St Omer—splendid young fellows who, at their country's call, were preparing themselves there for a temporary commission, and who saw clearly that in the fighting line they would be lucky if they escaped wounds or death for more than a week. And as the casualty lists mounted, the people of Britain found themselves united, as never before in their history, in a nation-wide fellowship of bereavement and sorrow.

Among the fighting men the initial enthusiasm was now tempered by a grim realization of the price that had to be paid. Some of the more sensitive spirits have told in deeply moving prose and verse of the horrors of life in the trenches. Some of the poetry of the period (I recall more particularly the work of Siegfried Sassoon, Edmund Blunden and Wilfred Owen) has become part of our national literary heritage; with its distinctive notes of anguish and compassion, disillusionment and revolt, it will continue to be read down the years in condemnation of the barbarism and wastefulness of modern war.

Yet there is another side to the terrible story; and this too must be kept in mind if we are to appreciate fully the temper of the time. The old 'death or glory' spirit had indeed largely gone; but in its place there was now in all ranks a quiet fortitude and a resolute determination to carry on to the end. In the course of the battle I had some memorable talks, notably with young regimental officers, which left me in no doubt that for them something of the old idealism was still a powerful motive-force, and that, hateful as war was, there could be no question of turning aside at this stage.

Here we come to the real tragedy of the Somme—not just the bloodshed and the horror, but the fact that to so many of those engaged in it no other way seemed open; and they accepted it, not in elation, but in a spirit of dedication. This, for example, is how one young officer expressed himself in a farewell message to his parents: 'Do not mourn for me, for I have laid down my life in the noblest cause the world has ever known, and feel I have not shirked my duty.' And there were countless farewell messages breathing the same spirit. As a signal illustration of what I have in mind I recall the story of the Rev. Gavin Pagan, whose name will continue to be held in honour by all who were privileged to know him. One of the leading ministers of the Church of Scotland (he was minister of St George's Parish Church in Edinburgh), and already in his forties when war began, Pagan had felt impelled to enlist as a private in the Royal Scots, and he was now in France as an officer with his battalion. Haig had heard of his story, and approved of the suggestion, strongly urged from many quarters, that a man of such character and experience might well find better use for his distinctive gifts in the work of an army chaplain. But Pagan himself would have none of it. Violence and bloodshed were abhorrent to him; but he was a man of earnest purpose and calm judgment, and he saw his duty clear. Before the Somme battle began I tracked him out in his tent at the base camp of Etaples; he had the firm set face of a hero of ancient Rome, but in his eye there was also the light of Christian hope. He was preparing then, as I recall, to go out on a night exercise. And in a sense that was typical of him. For the whole war had become for him a night exercise in which before it ended he was to lay down his life. But, as he saw it, it was a night exercise from which there was no escape, and it must be carried out without faltering if the new day for which he hoped and prayed was to dawn for his country and for humanity.

Such idealism is not readily appreciated by a generation that has grown weary of war, and recoils from the very thought of it. If it was war-fever, it was the war-fever of peace-loving men, fighting and dying, as they dared to believe, so as to end war for ever. Rightly or wrongly German militarism was seen not merely as a threat to British national interests, but as an offence to the conscience of all who cared greatly for the basis of Christian civilization. And so, though Germany began about this time to make tentative approaches towards ending the war, there was little disposition in Britain to welcome them. The nation shrank from the thought of a patched-up peace. Though the losses might become more and more severe, the war must go on.

There were grave questionings, however, in all the countries regarding the future conduct of the war. Both Germany and France made far-reaching military changes—Hindenburg and Ludendorff had taken over (after Verdun) the control of German operations in the West, and now in France both Joffre and Foch lost their commands. Joffre's post as Commander-in-Chief was given to the self-confident and impetuous Nivelle, who, having gained a striking local success in the fierce fighting around Verdun, held out before the French Government the specious hope of a speedy break-through which would roll up the whole German line. No change was made, or even seriously considered, in the command of the British forces. But Asquith, as British Prime Minister, gave place to the more vigorous and imaginative Lloyd George, who had previously served as Minister of Munitions and then as Secretary for War. Lloyd George made certain welcome changes in the composition of the War Cabinet, but he began too, to press his own ideas with regard to military policy. He had been deeply perturbed by the British losses on the Somme, where the situation, as he interpreted it, pointed to stalemate. Any continuance of the offensive in the West was bound, he felt, to be costly and ineffective. He accordingly pinned his faith to finding some round-about and less expensive way of compassing the enemy's downfall, perhaps by an advance in the Middle East or an all-out attack from Italy against Austria.

How did Haig view the situation? His military instincts told him that a German army massed so near our shores had to be met where it stood; there was no escaping that issue.

D

Further, he did not for a moment believe that the position was stalemate. What the Somme battle had accomplished fell very far short in many respects of what he had hoped for, and he was deeply conscious, though outwardly he did not show it, of the losses it had involved. But one thing it had accomplished, and that one thing was all-important. It had effectively destroyed the myth of German invincibility. The proud German army had been forced to recognize that it was not all-powerful; and Britain's citizen-soldiers had proved to themselves, and to the world, that they were more than a match for it. Haig was prepared therefore to face the future with a quiet confidence. His one anxiety was that the Allies might be stampeded into rash and ill-considered policies when the hour called for wise judgment, a steady nerve, and invincible faith.

He would sometimes give me his views on those matters, talking calmly and dispassionately as always, but with an unmistakeable note of conviction; and as he talked I came to appreciate how the outlook of the trained military mind may differ from that of the civilian. Keeping steadily in view the ultimate goal, the soldier schools himself not to be unduly perturbed by the day-by-day happenings; and there are occasions when progress is to be assessed less by the amount of ground gained than by the weakening of the enemy's power of resistance. On an over-all view of the Somme battle Haig did not doubt that it was a notable victory. Apart from relieving the pressure on Verdun it had forced the German army and its commanders to face the possibility of defeat. How right he was in this judgment was later to be confirmed from enemy sources. Hindenburg's conclusion was emphatic: 'The men must be saved from a second Somme.'

And what of the casualties? In a war of this magnitude, in which Britain was being called to bear an ever-increasing burden, Haig knew that casualties were bound to be heavy. If, with Mr Lloyd George, he could have persuaded himself that victory could be achieved in some circuitous and less costly way, how gladly would he have welcomed it. But as a realist he was convinced that there was no such way; and to talk and act as if there were was sheer escapism.

Nevertheless as a man Haig felt keenly the losses and the sufferings of the men in the trenches. He did not speak much about this. But no one who knew him had any doubts that he was deeply and personally concerned. Now and then his deeper feelings would come to the surface. For example, in a letter to Lady Haig in April, 1917, he confessed to 'a tremendous affection for those fine fellows who are ready to give their lives for the Old Country at any moment'; and he added: 'I feel quite sad at times when I see them march past me, knowing as I do how many must pay the full penalty before we can have peace.' That sense of 'sadness' was indeed never far absent; we find it coming to expression more than once in his private papers.

There is one notable passage in his diary which I must not omit to quote in this connection; it expresses the mingled pride and distress which he felt when in March, 1917, following on the German strategic retreat, he visited the scenes of the worst Somme fighting. He begins by saying that no one can visit the Somme battlefield without being impressed with the magnitude of the effort made by the British army. He goes on to emphasize how 'credit must be paid, not only to the private soldier in the ranks, but also to those splendid young officers who commanded platoons, companies and battalions'. Then come these significant words: 'To many it meant certain death, and all must have known that before they started. Surely it was the knowledge of the great stake at issue, the existence of England as a free nation, that nerved them for such heroic deeds. I have not the time to put down all the thoughts which rush into my mind when I think of all those fine fellows, who either have given their lives for their country or have been maimed in its service.' A man who could write such words was clearly not callous or indifferent. They were written in pride; but it is the pride which a leader who is by nature modest and sympathetic has in the troops whom he has been privileged to command. There is indeed not a word about himself in the whole passage until one comes to that closing sentence, in which he gives us a glimpse of how he felt about the sufferings and death of his men.

Those of us who saw much of the Chief at this time often remarked how well he seemed to have stood the strain. He never seemed over burdened or over anxious. Rather he had the look of the Happy Warrior,

> 'Who, if he be called upon to face
> Some awful moment to which Heaven has joined
> Great issues, good or bad for human kind,
> Is happy as a lover.'

In the midst of the uncertainties of the period, he received, as the year 1916 was drawing to a close, a letter from His Majesty the King which was a source of great encouragement to him. It reads: 'I have decided to appoint you a Field Marshal in my Army. By your conspicuous services you have fully merited this great position. I know this will be welcomed by the whole Army in France, whose confidence you have won. I hope you will look upon it as a New Year's gift from myself and the country.'

CHAPTER 7

1917 (i):
TENSIONS AND FRUSTRATIONS

1917 was likely to prove a decisive year; it might not bring an end to hostilities, but it would surely prepare the way for a decision in 1918. Time was now seen to be an urgent factor; war on such a scale, and pursued with such intensity, could not be continued much longer. But the deadlock on the western front remained unbroken. How then was a decision to be reached? A rash move in the wrong direction might well bring disaster.

Events soon took a menacing turn for the Allies. In Russia growing disaffection led in time to revolution. Germany began to use her submarines to sink neutral as well as allied shipping, pursuing the policy with such ruthlessness that the British First Sea Lord, Admiral Jellicoe, reported pessimistically to the War Cabinet in June: 'There is no good discussing plans for next spring; we cannot go on.' This pirate policy had one other consequence: in April the United States of America entered the war on the allied side. But it would be 1918 before American aid in the field became fully effective; would the enemy succeed in bringing the Allies to their knees before then?

To those of us who lived through it, as to those whose viewpoint is that of a later generation, 1917 stands out as one of the blackest years of the present century. It was a year of frustration and tragedy. And it has become customary in many quarters to connect Haig's name with the tragedy—the tragedy which culminated in Passchendaele. We ought to remember that for Haig it was a year of frustration, when his own plans were foiled by the enforced adoption of a rival policy that nearly lost us the war.

Convinced that the Allies were slowly but surely gaining the upper hand, Haig regarded it as imperative that they should pursue a well-based over-all strategy, and refuse to be diverted from it; and his own plan, prepared in full agreement with Marshal Joffre, had been to launch in the spring a strong

attack in the Flanders area, designed to clear the Belgian coast
and to deprive the enemy of her submarine bases there. Unfortunately there were leaders in both France and Britain who,
lacking his military insight and strength of purpose, decided
that a more adventurous policy might produce an earlier and
less costly victory. The French Government under M. Briand
endorsed General Nivelle's grandiose scheme for securing a
spectacular victory on the French front which would redound
in special measure to the glory of France; and Haig had
reluctantly to set aside his own plan of campaign so as to
cooperate by a subsidiary attack on the Arras front.

The British Commander-in-Chief was to be frustrated in a
more subtle and sinister way by the attitude of his own Prime
Minister. With all his zest for victory Mr Lloyd George was
never reconciled to the cost that victory over Germany was
bound to exact. Strongly opposed as he was to further British
offensives on the western front, he suddenly became an
enthusiastic supporter of Nivelle's projected offensive; and at
the notorious Franco-British Conference at Calais in February
he even set himself (without previous warning to Haig or Sir
William Robertson, who were present) to secure agreement to
a secretly prepared document, according to which the British
forces in France were now to come directly under the orders
of Nivelle, and the British Commander-in-Chief with his GHQ
staff were for all practical purposes to disappear.

It is important to recall this conference (rightly described[1] as
'one of the most unsavoury episodes in British political
military operations') because of the light which it sheds on
Haig's qualities of character. Lord Hankey,[2] who was present
as Secretary to the British War Cabinet, has told of the conster-
nation with which he read this document: 'it fairly took my
breath away'; and it may be imagined what a shock it was for
Haig. He could not but see in it an outrageous personal insult to
himself; but what disturbed him far more was the insult to the
British army which he had been called to command, and the
dangerous effect which such hasty and ill considered action

---

[1] Captain Cyril Falls, *The First World War*, p. 249.
[2] Lord Hankey, *The Supreme Command, 1914–1918* (London, Allen &
Unwin, 1961), II, p. 616.

would inevitably have on the chances of success in the coming offensive and on the whole cause of inter-allied harmony and cooperation. A lesser man would almost certainly have been tempted to offer his resignation at once. But Haig, with his unfailing sense of great issues, realized that, however welcome it might be to the Prime Minister, such action on his part at this critical hour would create disunity and dismay throughout both nation and army, and do serious harm to the Allied cause. Haig was ready, as always, to sink purely personal considerations in devotion to the cause which he served.

Fortunately it soon became evident that the original document was altogether impossible; a new formula was reached which Haig, though with some reluctance, was prepared to accept; and it was due not a little to his magnanimity that explosive developments were avoided and that the discreditable story for long remained a secret. But the experience left its mark on him. He was appalled by the underhand behaviour of the Prime Minister, and was never afterwards able completely to trust him. Nivelle too began by being impossibly tactless and domineering; and as a relief to his feelings Haig wrote in his diary: 'It is too sad at this critical time to have to fight with one's Allies and the Home Government, in addition to the enemy in the Field.' But such a thought is not one on which Haig would have allowed himself to dwell for long, and I question whether he would ever have given expression to it in conversation even to an intimate friend. There was no place in Haig's nature for pique or self-pity. He rode the storm with a serene self-possession; and he emerged from it a greater man than before.

The test came soon. On April 9th (Easter Monday) Haig opened the British attack on the Arras front; and including as it did the capture of the redoubtable Vimy Ridge this was by far the most striking success which the British troops had achieved since the war began. Nivelle's much vaunted offensive followed a week later; and it proved from the first so costly a failure that it had to be called off. And with its abandonment went the collapse of the whole plan of campaign which, on French initiative, the Allies had so rashly and unwisely adopted for 1917.

The shock to French hopes, and to French pride, was almost overwhelming. Alarm, despondency, suspicion and distrust began to spread throughout the country and among the troops. Acts of mutiny occurred on an alarming scale. Nivelle was dismissed and passed from the scene. His successor, General Pétain, devoted the first months of his command to the arduous task of enforcing discipline and restoring morale. But for months to come the French army could no longer be relied on as a fighting force. The summer of 1917 was critical for the whole issue of the campaign; how critical it was is apt today to be forgotten. The situation was in some ways parallel to that in 1940. Fortunately in other respects it was radically different. Had France in 1917 lost heart, and allowed the enemy to break through, the war was as good as lost. It is to the credit of France that this did not happen. But her recovery was to be long, slow, and at times uncertain; and it was only made possible by the stability, resolution and sacrifices of her British ally, who now had to bear the main burden of the fight just as France had borne it in 1914 and 1915.

In particular the British Commander-in-Chief was ready to do everything within his power to give France the encouragement and help she so much needed, and so prepare the way for victory in 1918. During the early summer of 1917 he emerged as the dominant military personality on the western front. While all around there was a tendency to doubt and despondency, to recriminations and divided counsels, here was a man whose one aim was victory, who had a well-grounded confidence in victory, and in pursuit of victory was ready to be wholly cooperative, wholly disinterested.

During those anxious and arduous months in 1917 my relations with Haig, always pleasant, became much more intimate and personal. On Easter Sunday, the day before he launched his big offensive on the Arras front, he had attended church as usual at Montreuil; and in his diary he noted that one of the prayers at the service had been for 'an unconquerable mind'. It was characteristic of his thoughtfulness that, before he left that afternoon for the forward zone, I received a letter to tell me about the coming operations, to say that, if I cared, I might join some of the Scottish troops who were to be engaged. I was thus free to spend the next three weeks with the troops in and around Arras, until I was recalled in the end of April to an Advanced GHQ which had been established in a rather squalid little village, Bavincourt, on the road between Arras and Doullens. There we were a very small company indeed—only a few senior members of certain staff departments, with the necessary complement of clerks, orderlies and troops for sentry duty; and during our four weeks' stay there I was the only resident chaplain, free (apart from Sunday duty) to spend most of my time nearer the fighting line.

Quarters had been found for the Field-Marshal in an unpretentious chateau close to the roadside; and on arriving at Bavincourt I went at once to call on him, and heard from a member of his personal staff something of recent developments following on the collapse of the Nivelle offensive. At the urgent request of the French Government Haig had been absent for some days in Paris, where the leaders were acutely disturbed by the course of events. Both the eighty years old M. Ribot, who had displaced M. Briand as Premier, and his very able War Minister, M. Painlevé, were most eager to have Haig's advice on a situation which threatened to get completely out of hand. Mr Duff Cooper in his Life of Haig had commented on what he calls 'the irony of the situation'. 'Two months earlier' he writes, 'the French Government, with the enthusiastic support of the British Prime Minister, had been seeking to compel the British Commander-in-Chief to become a mere automaton under the inspired guidance of his more gifted French colleague. Now the French Minister of War was almost on his knees to that same British Commander-in-Chief to furnish him with material that might help him to get rid of that same French colleague.' It is right to add, for the light it sheds on Haig's character, that the British Commander was careful to say nothing in Paris that might lessen the French Government's confidence in Nivelle, and indeed he would have preferred that at so critical a time there should be no change in the French High Command.

In June a new Advanced GHQ, on a much larger scale, was established at Blendecques, a picturesque village near St Omer;

and this continued to be the main advanced headquarters till December. Haig was now free to revert to what had been his original aim, the clearing of the Belgian coast; and after the capture of the redoubtable Messines Ridge, a success even more spectacular than the capture of Vimy Ridge in April, he pushed on with his plan for an attack beyond Ypres, in which it was hoped that, if sufficient progress were made, the Navy might cooperate by landing a force behind the German lines.

Returning as I did each weekend to Advanced GHQ, I generally lunched or dined on Sunday with the Chief, and so met many distinguished guests—army commanders, members of the Government, and others. Among the army commanders I met then were Byng, the victor of Vimy Ridge, a man of great charm and rich humanity; Gough, the youngest of them all (he was forty-seven), full of vigour and dash, who was to play a major part in the coming battle; and Rawlinson, whom Haig valued greatly not merely for his military gifts but for his unfailing good spirits. Lord Derby, the Secretary of State for War in this critical year, was a frequent visitor. A staunch supporter of Haig, he had his difficulties in plenty with the Prime Minister; and Haig in turn attached great value to his loyalty and straightforwardness.[1]

One guest whom it was a special privilege to meet was Viscount Esher, who at that time held a semi-official position in Paris, performing 'behind the scenes' a service of incalculable value to the British Government in the all-important field of inter-allied relationships. He was a frequent visitor to GHQ; if he was there on a Sunday morning he always came to church with the Chief, and I was generally invited afterwards to meet him. Esher was a man of rare distinction. He had been selected after the South African War to preside over a Commission for the Reorganization of the Army; in 1905 he had become a permanent member of the Committee of Imperial Defence; and he had continued ever since to be a firm believer in Haig's outstanding military ability. He moved freely in exalted circles, knowing (as was said) everyone worth knowing; but he de-

[1] A rather petulant entry in the Diary (January 14, 1918) is not, I am sure, to be taken too seriously.

clined all invitations to accept public office, preferring to exercise his great influence on his country's behalf through unobtrusive but pervasive channels. I have warm recollections of the many kindnesses I received from him, more particularly of one long and intimate conversation as we sat together in August 1917 in the garden of the chateau at Blendecques, and of visits I was privileged to pay to him soon afterwards in Paris and at his Scottish home, the Roman Camp, near Callander in Perthshire. If I recall all this now it is because of the fresh light I gained from these talks on the critical war situation—not least on the demoralization in the French army and intrigues in the French capital, the uncertain and conflicting policies of the British Government, and the confidence with which the British people still looked to Haig to achieve final victory in the field.

As the summer of 1917 advanced I could see that it was a time of growing anxiety for Haig. In his talks with me he often relieved his feelings with regard to our French allies. They had suffered much, no doubt, but 'what was one to make of an ally that was elated when all went well, but ready to lose hope when things went ill?' I would add, however, that he never showed any lack of confidence in the ultimate outcome. Indeed, I had occasion frequently to note that there seemed to be a new alertness in his step, a more buoyant glance in his eye, as if with increasing responsibilities he was rising to new heights of self-assurance. The time had come, as he saw with unerring instinct, when the British army must be prepared to meet with only limited allied support the main German challenge in the West, and meet it by a vigorous and confident offensive; only in this way could we wear down German strength, restore France's morale and reawaken her belief in final victory. And in so far as this challenge came to himself as Commander-in-Chief, Haig had no hesitation in accepting it. It added greatly to his anxieties, however, that the Prime Minister at home did not share his confidence, preferring that our policy in the West should be mainly defensive until America should be ready to put an army in the field. Haig gravely doubted whether France would be prepared to wait and suffer for another year; and if France failed at this stage, the war might well be lost.

I shall always recall with special interest the church services at Blendecques.[1] At some other places the smallness of the congregation made the conduct of the service far from easy; it limited me greatly, for example, in my choice of suitable hymns. Lord Northcliffe has told how,[2] at a service he attended in the course of the 1916 Somme battle, the singing of an unfamiliar hymn was so half-hearted that I intervened after the second verse and gave the number of 'Rock of Ages', which (he says) was 'sung with intense devotion'. The Press Lord on that occasion was obviously interested in our little hymn-book, With the Colours, for holding a copy in his hand he approached me at the close of the service with a solemn air, saying: 'May I commit a theft, and take this with me?' I refrained from telling him that on a previous Sunday Sir William Robertson had evidently been similarly moved, but had felt no need to ask permission before putting a copy in his pocket. By contrast with some earlier experiences I now generally had at Blendecques fifty to one hundred worshippers, including visitors from hospitals and army units in the near-by town of St Omer; and the gravity of the military situation served to deepen the sense of worship.

Sunday, July 22nd, was a day that is full of memories for me. Preparations were proceeding apace for the big offensive that was due to open soon in the Ypres sector; and on that very lovely summer morning it was hard to contemplate the riot of bloodshed that might break out before another week had passed. At the church service[3] I had preached on Hope—not the hopefulness of the natural man, which is often a mere matter of temperament, and apt to degenerate into wishful or presumptuous thinking, but the well-based hope of the Christian believer which is essentially a moral asset, the fruit of a strong hold on Christian truth and a rich spiritual experience, enabling him even in the darkest hour to go forward with confidence and faith.

I dined that evening at the Chateau; and what a company

---

[1] There is a description of one of these services in F. S. Oliver's posthumous volume, The Anvil of War, p. 263.

[2] Northcliffe, by R. Pound and G. Harmsworth, p. 505.

[3] General Charteris refers to this service in his published Diary, At G.H.Q., p. 235.

George Duncan in the uniform of a chaplain fourth class, the lowest grade in the Army Chaplains' Department. British Army chaplains did (and do) not use standard ranks; a fourth class chaplain wears three stars, as for a captain.

General Haig in 1916; he was promoted to field marshal in January 1917 in the King's New Year Honours List.

George Duncan
during the war in
relaxed mode.

The Scottish Churches Hut at Montreuil sur Mer. The photograph can be dated no earlier than 1917, as only in that year were ladies allowed into the GHQ area, including three 'Scottish ladies', who assisted Duncan in the running of the hut.

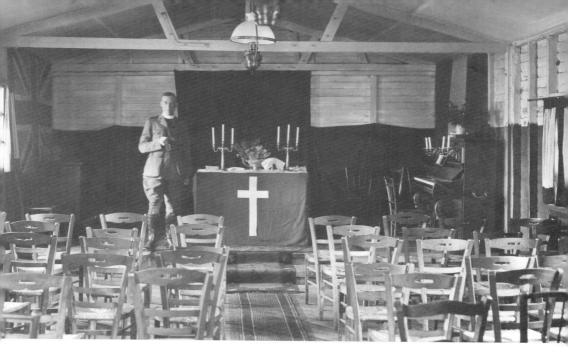

The Scottish Church Hut prepared for a service. Haig was a regular (but not necessarily weekly) attendee at the services here when based at GHQ.

Haig rode regularly – preferably every day – not only out of enjoyment, but for exercise.

Duncan, drawn in 1918.
To his left is the Abbatial
Church of Saint Saulve in
Montreuil, which possesses
a magnificent late gothic,
west entrance.

Haig looks on as men enter the courtyard of the École Militaire, a building in Montreuil used by various departments of the GHQ, for a special service of Thanksgiving on 4 August 1918.

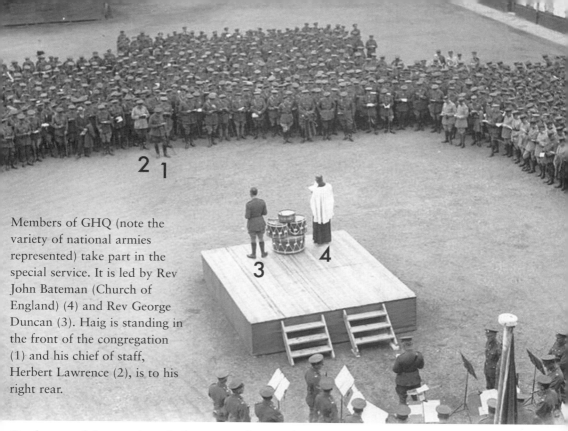

2 1

Members of GHQ (note the variety of national armies represented) take part in the special service. It is led by Rev John Bateman (Church of England) (4) and Rev George Duncan (3). Haig is standing in the front of the congregation (1) and his chief of staff, Herbert Lawrence (2), is to his right rear.

3 4

On the eve of the opening of what was to become known as 'The Advance to Victory' or 'The Hundred Days', Haig poses with the King, President Poincaré and some French generals at the entrance to Haig's residence, Chateau Beaurepaire, near GHQ in Montreuil.

On the same visit, George V is seen with with Haig and the
Prince of Wales walking by a guard of honour.

Haig greets the commanding officer of the Nova Scotia Highlanders (CEF), near Domart-sur-la-Luce on 11 August 1918, a few days after the opening of the Battle of Amiens on 8 August ('the Black Day of the German Army'), in which the Canadian Corps played a prominent part.

Haig passing another Canadian Scottish battalion, also near Domart, a village that had been almost completely destroyed in the fighting of 1918.

When the Armistice came into effect, Haig was based on his train, which he used as an advanced headquarters. Here he is greeting a Japanese delegation, led by Prince Fushimi of Japan, a member of the Japanese royal family who probably had the closest connection between it and the British royal family; he represented Japan at both Edward VII's funeral and George V's coronation. Nearest the carriage door behind Haig is Philip Sassoon, his private secretary (and cousin of the author-poet Siegfried). The Prince of Wales is on the right of the photograph, talking to Fushimi.

On a busy day, Haig was photographed on the steps of 40 Place Eugène Thomas, Cambrai, surrounded by his army commanders. From left to right: Plumer (Second Army), Byng (Third Army), Birdwood (Fifth Army), Rawlinson (Fourth Army), and Horne (First Army).

Earl and Countess Haig shortly after his elevation to the peerage in March 1919, with their two oldest children, Alexandria ('Xandria') and Victoria Doris ('Doria').

The family portrait taken on the baptism of Haig's last born child, Irene ('Rene'), on 11 November 1919. His son, Dawyck, is firmly in the grip of Doria!

Haig enjoyed playing golf when time and opportunity allowed; Combe Hill Golf Club, October 1919.

In his final years, Haig devoted much time to veterans' welfare.

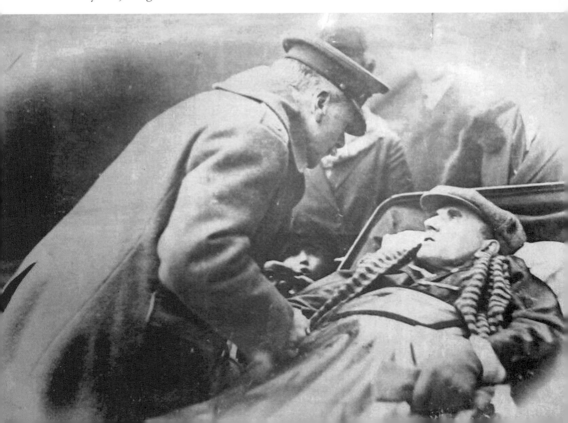

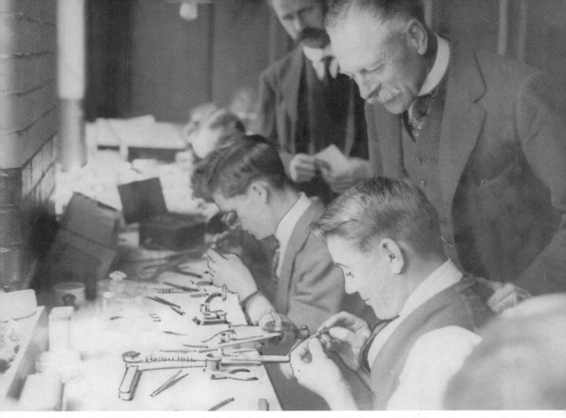

Haig visiting the Willesden Polytechnic Training Centre for Discharged Soldiers (now the College of North West London) on 9 June 1920. The men on this particular course are being trained in watch repairing.

Lord and Lady Haig made very successful post-war tours to South Africa (1921) and Canada (1925). Here he is photographed playing deck quoits on the deck of the newly launched SS *Letitia* in April 1925.

The tour of Canada involved extensive travelling by train – and allowed for photographs of a relaxed field marshal.

Also in Canada, where he reviewed a great parade of veterans, a rare photograph of a smiling Haig in uniform.

In Alberta in late July, Haig tried his hand at golf on the recently constructed Jasper Park course, which apparently took 50 teams of horses and 200 men a year to construct it. Haig is seen greeting one of the 'cowboys' who helped to build the course.

Haig with Irene in 1927 at the home provided by public subscription in the Borders, Bemersyde. Although active in support of veterans' organisations, Haig enjoyed the life of an active landowner.

Lord Haig making a speech near Sir Dhunjibhoy Bomanji's (to the right rear of Haig) estate at Windsor. The occasion is the presentation of the latter's donation of £5,000 to HRH Princess Alice for the construction of a Comrades of the Great War Club at Windsor. This organisation merged with the British Legion in 1921, not least because Haig was keen that ex-servicemen's associations should unite and speak as one force, rather than have a scattered and therefore relatively weak voice in public matters. Amongst those present (or at the subsequent march past in Windsor) were 300 war orphans, war widows and veterans, disabled and fit, including Windsor's two VC winners. Sir Dhunjibhoy was a wealthy shipping magnate and a great admirer of Haig. He commissioned the equestrian statue of Haig that stands in Edinburgh Castle, which he presented to the corporation in 1923.

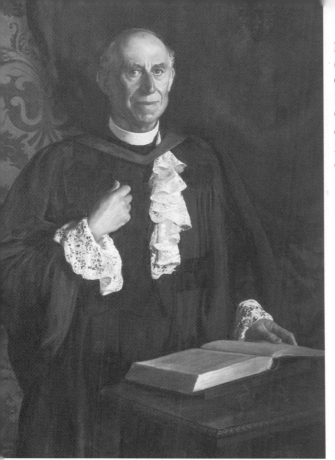

Reverend Professor George Duncan (1884-1965), Professor of Biblical Criticism, St Andrew's University 1919-1954, Principal of St Mary's College 1940-1954, Vice Chancellor of the University 1952-1953. Moderator of the General Assembly of the Church of Scotland 1949.

Haig's funeral was a considerable public event, with the coffin lying in state in St Columba's, London, before being moved in a ceremonial procession to Westminster Abbey. In this photograph taken during the procession, his horse, with boots reversed, is preceded by Haig's batman of twenty-five years, Sergeant Thomas Secrett. Afterwards the coffin was moved to St Giles Cathedral in Edinburgh, where it lay in state for three days before a private burial at Dryburgh Abbey, a short distance from Bemersyde.

Sir James Guthrie's (1859-1930) portrait of Earl Haig (1861-1929), 1923.

I found assembled there. Here was General Pershing, who was to command the American Forces in France, and with him his CGS, Lieut. Col. Harbord; having quite recently arrived in France they were now paying their first official visit to the British Commander-in-Chief. Here too was Sir William Robertson (Chief of the Imperial General Staff), who had arrived in time for dinner, having crossed from England in a destroyer; and there were also present three of our own GHQ generals, Kiggell (CGS), Butler (Deputy CGS) and Charteris (Chief of Intelligence). Robertson I had met on previous occasions. I could not but view with admiration this man who had risen from being a trooper in the Lancers to be the Government's chief military adviser. I knew, too, something of his problems vis-à-vis the Prime Minister, for he was a convinced believer in the vital importance of the western front and strongly opposed to any diversion that would seriously weaken our position there. Sitting next to him at dinner that evening I found conversation with him easy and natural. He confessed to a chafing of spirit in his present tasks, having (as he said) constantly to deal with politicians in London instead of with the enemy in the field. 'But then,' he added, 'we must all learn to do what we are given to do, and my job is there just as yours is here.'

It was a lovely summer evening; and after dinner we went out to the garden, where a long table was set under the window of the Chief's working-room. Almost at once Haig came up to me, saying, 'You must have a talk with General Pershing'; and, with his usual courtesy, he insisted on Pershing and me being seated in the only two available camp-chairs, while he contented himself with a less comfortable seat. I recall how Pershing impressed me with his quiet confidence and strength; his manner was reserved and unassuming; he seemed to me to be made in much the same mould as Haig, though lacking Haig's wealth of experience and in that indefinable 'something' which served always to set Haig apart from his fellows. He went on to tell me how he and his CGS had spent the last week visiting various army units to familiarize themselves with the British war effort; and he confessed that its magnitude had come as a complete revelation to him.

Still sitting in the garden I had quite a long talk after this with the Chief. Among other things he told me that he had had the Archbishop of York (Dr Cosmo Lang) to lunch that day, and that he had been much drawn to him. 'We had been at Oxford together; he went up about the time when I went down. We talked of changes that would have to be made in the Church after the war; and on this matter he seemed open-minded and sympathetic.' The Chief then told me that he had spoken very frankly to the Archbishop about the reasons why he now always attended the Church of Scotland service. 'I told him how disappointed I had been with some C. of E. services I had attended. Chaplains ought to deal seriously with the great issues confronting us, and do so in a way that people understand.'

The Chief had been in splendid form the whole evening—calm and restrained, as he always was, but in his own quiet way happy and even hearty, pleased to be among friends, and attentive and courteous towards them as ever. I quote some words I wrote about him in my diary that night. 'He has, I am sure, aged quite considerably in the last eighteen months. But there is never a burdened look on him. I have often seen him with his eye flashing with the joy of action for which he was ready; tonight (as on so many recent occasions) there was rather a steady, straight look on his face, as if he had much on which to meditate before the action began. But he never seems to be oppressed by his meditation. There is nothing self-centred about D. H., nothing that is moodily self-conscious. His thoughts are centred on the work that has to be done.'

Despite the ease with which he chatted to me and others that evening, he had certainly much to cause him anxiety, more especially in regard to the attitude of the Government. An official communication, received two days before from Sir William Robertson as CIGS, had conveyed the Government's belated approval for the Flanders attack, but asked, too, for arrangements to be made to send troops to Italy, if the attack did not immediately succeed. Haig's diary reveals how profoundly this communication had distressed him. Convinced as he was that we were standing now at the crisis of the war, the thought of diverting troops at such a time to a subordinate theatre seemed to him to be utter lunacy. His distress was increased by Robertson's frank revelation in a personal letter that had followed the above official communication. 'We had a rough and terrible meeting yesterday,' Robertson had written.[1] 'The fact is that the Prime Minister is still very averse from your offensive, and talks as if he is hoping to switch off to Italy within a day or two after you begin.'

It is not surprising, therefore, to learn that, on going indoors after his talk with me, Haig had a very frank discussion with Robertson on the immediate situation and on Government policy. After handing over a carefully worded reply to the official communication of two days before, he spoke to Robertson with more than usual seriousness on the need for a sound strategical policy, urging the CIGS to stand firm and not acquiesce in the setting aside of the clearly expressed advice of the General Staff.

There is an addition I should like to make to this picture of Haig on that memorable Sunday. At some time in the course of the day he had sat down at his desk to prepare a document which may still be read among his private papers. At its head stands the word 'Notes' with the date (July 22, 1917); and in ten short paragraphs he stated his views on the critical war situation, and on the aims which ought to govern all our strategy. For clarity of thought and succinctness of expression the document is indeed a masterpiece. It deals with 'the weariness and disappointment of the French' and our obligation meanwhile to safeguard and encourage them, and so nurse them through the winter; for it is practically certain (he adds) that next year's campaign will be 'the last we shall induce France or Russia to face (even if we can get them to face that, which is doubtful)'. To ensure success next year we must concentrate our efforts now and not dissipate them. And his summing-up is as follows: 'Every man and gun that we send to Italy reduces our power in one respect or the other. I consider it so dangerous and so unsound to adopt such a course that, in

[1] See Robertson's Memoirs, *Soldiers and Statesmen, 1914–1918*, vol. ii, p. 248.

my opinion, any responsible soldier who consents to issue an order for it must expect to be adjudged by history to have failed in his duty to his country.'

We do not know what object Haig had in writing this document. But the reference to 'any responsible soldier' in the concluding paragraph suggests that it was written in preparation for the talk that followed that evening with the CIGS.

CHAPTER 8

1917 (ii): BAFFLED HOPES: PASSCHENDAELE AND CAMBRAI

After two enforced delays there opened on July 31st one of the grimmest struggles in the history of war, the Third Battle of Ypres, often but incorrectly referred to (from the name of the village and ridge which became the final objective) as Passchendaele. Haig's hope had been that sufficient progress would be made at the outset to allow for naval cooperation, a landing from the sea, the clearing of the Belgian coast, and an advance towards the important German rail-centre of Roulers. As it was, the initial attack achieved only a limited success; lashing rain began to fall in the afternoon and continued for three days; and soon the shell-torn ground was turned into a quagmire. Fighting continued intermittently during August and September; and when October brought a renewal of persistent rain, conditions became something of a nightmare. Haig, however, was loath to end operations until the capture of the whole of Passchendaele Ridge should provide a drier and more secure line for the winter. With the capture of the ridge on November 6th the offensive was called off four days later. As for the immediate results, the advance was at no point more than 10,000 yards, and British losses reached the appalling figure of 240,000. On the other side of the reckoning, German losses were certainly not less, and were probably greater; and if the sufferings which our men endured so heroically were tragic beyond description, the experiences of the German troops were equally tragic, and seriously undermined their confidence and fighting spirit.

The offensive certainly fell very far short of expectations. And this lack of success, coupled with the terrible losses and suffering involved, has given the name Passchendaele an evil reputation. Some popular delusions, however, are now giving way before fuller study and calmer judgment. Captain Cyril Falls, in his recent book *The First World War*, dismisses the belief that it was all 'mere blind bashing'; and he adds: 'Tactics were seldom more skilful.' As for the idea, fostered by

E

Mr Lloyd George, that Passchendaele from first to last reflects the stupidity, obstinacy and callous indifference of the British Commander-in-Chief, such a contention is so at variance with facts that the serious thinker is impelled to seek a deeper explanation.

Meanwhile the debate continues; and Passchendaele provides the most controversial chapter in Haig's career. Was he justified in embarking on so doubtful a venture? Or in continuing the fight so long? Did he seriously underestimate the opposition to be encountered? Was he perhaps misled by the reports of his Intelligence Chief? These matters are still open for examination and discussion. On one matter, however, there is a growing consensus of responsible opinion : it is that for a true appreciation of Passchendaele we must set it in the context of the war as a whole, and try to view it from the standpoint of the man who had to make the decisions.

Haig was profoundly convinced that we now stood at the turning-point of the whole campaign. Germany's resources in the West were, he believed, approaching exhaustion; if we struck hard now, before she was able to withdraw her troops from the Russian front, it might even be possible to force a decision before the end of 1917. And the only power capable of such vigorous action was Britain. With France weak and demoralized it would be folly, it might even prove fatal, to allow the initiative in the West to pass into Germany's hands. To switch off divisions at such a critical hour to North Italy in the hope of possibly overrunning Austria was a gambler's policy, the results of which were bound to be doubtful, and might be disastrous. As Haig saw it, the Flanders attack was the one line of action clearly dictated by immediate needs; and if successful, it opened up wide possibilities.

I saw Haig at most week-ends during those anxious months. Acutely distressed as he must have been by the set-back to his plans and the mounting casualty lists, it was not his way to give open expression to his thoughts on such matters. I recall General Kiggell saying : 'Just look at him. He gives no sign of disappointment. But by God ! isn't he feeling it !'

Sometimes however he did express his thoughts with more than usual vigour and emotion. The lack of adequate French support at this critical hour caused him acute distress. Normally he was restrained in his criticisms of the French, and generous in his praise. But I think of one occasion at the close of September when the failure of the French stirred him to recall some past experiences with a vehemence which surprised me. He told me how, in the retreat from Mons in 1914, a French commander had tried to keep British troops and transports from using the only road that was open ('a fine way to treat your allies, isn't it?'), and how in 1915 the French command had incurred frightful losses in a series of attacks that could not possibly lead anywhere, 'This was fighting the enemy without intelligence.' 'I began,' he went on, 'with a readiness to believe the best of the French—as was only natural for a Scot, with memories of the old alliance; but events were too much for me.' As he spoke, one realized what it meant to him that, though nominally we had allies in Russia and France, Russia was now (by her own doings) almost out of the war, and France was not doing anything like what was called for; and at what was undoubtedly a decisive hour the British troops were being left to fight practically alone. I suggested to him that the situation recalled our earlier struggle with Napoleon—would Britain once again, by her exertions and example, be the saviour both of herself and of Europe?

It was, however, an essential part of Haig's greatness that disappointments, no matter how acute, were never allowed to come between him and concentration on the ultimate issue, which was victory. The name had cropped up at lunch one day of a divisional commander who in the course of the battle was showing himself restless and dispirited. At this the Chief turned to me (I was sitting next to him), and looking me straight in the eyes as if he expected me to understand and concur in what he was to say, he exclaimed: 'The fellow hasn't the faith to see that we can go through the enemy and beat them with bladders.' And it is a significant fact that so long as he lived Haig never wavered in his belief that the Flanders offensive, despite its heavy losses, had played an essential part in the winning of the war.[1] It had left the enemy

[1] Confirmation of this is provided by Hankey, *The Supreme Command*, II, p. 700.

battered and shaken; a demoralization had set in from which they were to find it hard to recover; certainly they would not endure another such hammering; and though they might, with strong reinforcements from the Russian front, attempt one last desperate offensive in the spring, there was still good hope of an allied victory in 1918.

That hope, as we know, was to be realized; within a year Germany was to ask terms for an armistice. And the question suggests itself: could the same result have been obtained in shorter time, and at less cost, if for those months we had adopted a defensive policy in the West, leaving Germany free to build up her strength or to mount an attack against the French?

With the close of the Passchendaele fighting the battle of the giants seemed to be over for another year. But Haig had other ideas. For many weeks he and his staff had been giving thought to a possible tank attack, swift and unexpected, in the hope of improving our strategic position before the winter set in, and increasing still further the enemy's sense of insecurity. Passchendaele, with all its failures and frustrations, had clearly not broken his spirit or sapped his confidence in final victory.

In approaching this story I recall how, on the evening of Sunday, November 18th, I was invited to dine with the Chief, and on my arrival I could see at once, from the members of the GHQ staff who were present, that something very unusual was afoot. Taking me aside before dinner, Colonel Alan Fletcher, the Assistant Military Secretary and senior ADC, put me au fait with the general situation. A surprise tank attack was to be made on Tuesday morning in the Cambrai sector. It had all been planned with the greatest secrecy; even the French command, he said, had not been informed of it until a few hours ago. And there was reason to think that the Germans had so far no anticipation of it. Night after night an immense force of tanks had been collected, and carefully screened, in Havrincourt Wood; and they would open the attack on Tuesday morning without a preliminary bombardment.

I was to learn more in the course of dinner conversation. On the previous night the enemy had made a raid (were they becoming suspicious?); and they had taken prisoner an Irish lad, who, however, was not likely to be able to give anything away as his Division was to be relieved before the attack. Later, before we parted for the night, the Chief led me into an adjoining room, where I was alone with him, beside his private desk and table and huge army maps that covered most of the walls.

After a few general words he pointed to a map on his table, and said : 'I know you are a man who can keep a secret. You will be interested to know more about this new attack; and you may want to go forward tomorrow so as to be on the spot when it begins.' He then gave me a general picture of what he had in mind. He had great hopes for tanks if they could be rightly used. But two conditions were essential—surprise and firm ground; and these two were likely to be fulfilled here. If the attack should involve the loss of a large number of tanks, he was prepared to accept that, for a new and lighter type of tank would be in supply by the spring. And he emphasized that any success that might be gained would be strictly limited; he did not have the troops to exploit it.

No one with experience of the Cambrai battle will forget the opening morning, November 20th. Without the least warning the stillness of the November dawn was rudely broken by an outburst of intense artillery fire; hundreds of tanks, followed by the infantry, emerged in the dim light, to crash a way through the wire entanglements; within a few hours first one line, then a second was breached in the Hindenburg defence system, and 6,000 prisoners were taken. But at two points there was a regrettable hold-up which was not rectified before darkness fell at four o'clock; and the delay led to fierce fighting, lasting for several days, for the control of the ridge at Bourlon Wood, the capture of which would have opened the way to Cambrai. The enemy, now heavily reinforced, were able to hold on to the ridge; and on the morning of the 30th they suddenly launched to the south a strong counter-attack which won back for them practically all that they had lost. We had come within an ace of a striking success, and had failed.

The Cambrai battle left the strategic position in the West practically unchanged. But I have recalled it here because of its far-reaching consequences. Taking a long view we can see how it inaugurated a new chapter in modern land warfare, providing a model for Haig's forward drive in 1918, and for the massed armoured attacks of the Second World War. Its more immediate effect was to centre attention on its failure. At home a mood of pessimism and distrust settled on the nation as this year of frustrated effort drew to a close. The Government too was rattled, and began to look for victims. Haig naturally did not escape criticism, though in his favour it was remembered how near the initial attack had come to being a quite astounding success, and much that went wrong in the course of the battle seemed to stem from a failure of some commanders in the field to act with due promptitude and decision. Strong pressure was, however, brought on Haig to make big changes in his GHQ staff. The more important of these concerned the replacement of Brigadier-General Charteris as Chief of the Intelligence Department, and of Major-General Kiggell as Chief of the General Staff.

Charteris's relations with Haig have occasioned so much controversy that his name calls here for more than passing mention. I knew him well. A man of abounding vitality, he tended to be bustling and boisterous; and his loud-voiced exuberance made him none the too popular with his immediate colleagues and with others up the line and at the War Office. It was also urged against him that, optimistic as he was by nature, he had in his reports to Haig repeatedly underestimated the enemy's resources and capacity for resistance.

In the circumstances of the time it was clear he had to go. But his faults ought not to blind us to his merits and achievements. It is of some interest to note that his family traditions were academic—his father, his two brothers and his father's brother were all university professors; and the last named of these (the Very Rev. A. H. Charteris), though not notably successful as a university teacher, displayed as a churchman such outstanding gifts of vision, initiative, and organizing ability that he has been described as a man who did more for the Church of Scotland than any other man since the Reformation.[1] To widen his outlook John Charteris had on leaving school

[1] See R. H. Fisher, *The Outside of the Inside*, p. 13.

spent a year in Germany before entering the army as a Sapper. Attracted by his ability when they were together in India, Haig attached him to his personal staff; and during the war he was Haig's Intelligence Officer with the First Corps, then with the First Army, and finally at GHQ. When he became Director of Military Intelligence he was the youngest Brigadier-General in the army.

C. E. Montague[2] of the *Manchester Guardian*, who as a war-correspondent was attached for a time to GHQ, describes Charteris as 'a fully educated man with a good fifty per cent more of brains, imagination, decision and initiative than the average of his fellow regulars on the Staff'. With his alert mind and youthful outlook he was alive to what was happening in other fields beyond those to which most military men confided their interest. In addition, he had a talent for organization. As a consequence he developed at GHQ and throughout the army an Intelligence Service which in many respects was far superior (so at least I often heard) to that of either the French or the Germans. Shortly after his departure I had a conversation with his immediate successor, Major-General Sir Herbert Lawrence; and he spoke to me with great appreciation of what was due to Charteris for a Service which was so efficiently organized and had such far-reaching ramifications.

If Charteris failed, it was not through lack of capacity. His faults were largely temperamental. In conversation with him I often reflected that his mind moved by leaps, and that some of his opinions (at least as he expressed them) did not fully reflect his considered judgment. Was the same perhaps true of his relations with Haig? Haig valued his work highly. The information he supplied on German troop movements was, I believe, remarkably detailed and accurate; much that he reported on Germany's internal difficulties and on the decline in her fighting spirit was perhaps not far from the truth. Where he seems to have erred, more especially in his face-to-face dealings with the Commander-in-Chief, was in the interpretation which he put, or allowed Haig to put, on the facts before them. Perhaps he had been too long in Haig's entourage, and suffered

---

[1] In his book, *Disenchantment*, chapter vii.

from lack of recent experience in the fighting-zone. With his departure criticism of our Intelligence ceased; a new spirit of harmony prevailed; but the three men, Generals Lawrence, Cox, and Clive, who in succession took his place in 1918 would all have acknowledged that they had the good fortune to operate a machine which Charteris had built up and set in motion.

General Kiggell, the Chief of the General Staff, was a less controversial figure than Charteris, but he too had lost the confidence of the Government. I had many talks with him; but he was a quiet, reticent, studious man whom it was not easy to know well, and even his name was little known throughout the army. Haig valued him for his gifts as a military thinker, and for his straightforwardness and reliability. But even at the best of times he seemed lacking in vigour, and by the end of 1917 he was obviously a tired man. To succeed him as CGS General Lawrence was promoted in January from the post he had held for only a few weeks as Chief of Intelligence; and there is no doubt that with his virile personality Lawrence proved in that last year of the war a more stimulating support for the Commander-in-Chief.

The year 1917 closed in gloom. It had been a year of foiled hopes; its heroisms and its sacrifices seemed all to have been vain. I recall the atmosphere of depression at GHQ, and a like depression prevailed at home. People were tired and bewildered. The war must continue; but for how much longer? And was it being waged on right lines? It was well for Britain that the two men most responsible for leadership, one at the head of the Government, and the other in command of the armies, were men who, however much they differed in other ways, still believed that victory was possible, and were determined to achieve it.

For the Prime Minister it was a time of very grave anxiety. For long dissatisfied with the military direction of the war, he was now firmly resolved to effect sweeping changes in policy and in personnel. He had been greatly elated by Allenby's victorious advance against the Turks, and the capture (in December) of Jerusalem; and he had high hopes that, by a bold imaginative strategy and with comparatively small loss to the attackers, the subsidiary enemy powers might all be 'knocked out' and Germany's position rendered hopeless. But however alluring the prospect, was it realistic? Did it view with sufficient seriousness the mounting threat to the western front? And was it likely that Germany would give up the struggle so long as her powerful army in the field remained undefeated?

Haig's anxieties too were very great—greater than at any earlier period of the war. The military prospect was never more gloomy. Memories of past failures and forebodings about the future might well have left him dispirited. To add to his anxieties it was only too plain that he no longer had the confidence and the support of the Prime Minister. And more was at stake in this than his own personal position. If there was to be a change in the command, there might also be a radical change in military policy, which at this critical juncture was bound to delay, and might possibly destroy, the hope of ultimate victory.

Mr Duff Cooper,[1] with his gift of words, has given us a picture of Haig in those closing months of 1917 that is worth recalling. 'Strong in the faith which others were losing, Haig steadfastly maintained his way, and it will be seen by how few that faith was shared and how solitary that way became. Until the very eve of victory, which still lay twelve months ahead, those doubts persisted in the minds of others and made them difficult colleagues, because they genuinely distrusted the soundness of his views. These were hard days for him, because he lacked both the power of exposition which might have proved his case, and the eloquence which might have inspired confidence in others. He was supported by the conviction that his strategical opinion, based upon the ceaseless study of a lifetime, could not be wrong, and by his religious faith which grew ever stronger with the passage of years.'

Haig has frequent references to the church service in his Diary at this time, and they help us to see how much his religious faith had come to mean for him. Charteris in his Life of Haig[2] recollects also a dull Sunday morning in December when I had based my sermon on Christ's prayer in

[1] *Haig*, ii, 163.
[2] p. 297.

Gethsemane: 'Father, if thou be willing, remove this cup from me; nevertheless not my will, but thine, be done,' followed by the answer recorded by St Luke: 'There appeared an angel unto him from heaven, strengthening him.' And he goes on to tell how the Chief, who as a rule was severely reticent on such matters, expressed his thought in conversation after the service: 'When things are difficult, there is no reason to be downhearted. We must do our best, and *for a certainty* a ministering angel will help.'

CHAPTER 9

1918 (i): BACKS TO THE WALL

A division between the campaigns of 1917 and 1918 is merely a matter of chronological convenience. From a military point of view the two campaigns must be seen as one. The death-grapple of 1917 continued unresolved. The errors and failures of that fateful year had still to be atoned for. The prospect of an Allied victory in 1918 had definitely receded. And with the collapse of Russia Germany now clearly held the initiative—would she attempt in the spring a decisive blow before American aid could become effective?

It was a grim outlook for the Allies. Fortunately a more resolute spirit now prevailed in France. A striking indication of this came when the aged national hero, Georges Clemenceau, nicknamed the Tiger, assumed office as Premier; and under his forceful leadership the French nation braced itself to meet every eventuality. In Britain there was no weakening of the national will; the nation responded to Mr Lloyd George's stimulating lead; but a sober realism now tempered the earlier optimism. How much longer must the struggle continue? And what further sacrifices must be faced before there was any prospect of ending it?

The war had become more and more a test of endurance. One might go further and say that it had become in no small measure a test of nerve. There was a growing readiness in many circles to believe that the German army was not to be beaten in the field, and that for another year at least its hold on the western front would remain unbroken. And among those who were disposed to acquiesce in this belief was the British Prime Minister who, for all his resolution, tended to look away from the western front, with all the losses that further fierce fighting there must entail, and to seek easier conquests elsewhere. His aim was to weaken Germany by 'knocking away the props'. In face of this situation Haig displayed a strength of character and conviction which from now onwards was to prove a decisive factor in achieving an early victory. Despite anxieties and frustrations that would have broken the spirit of most men he went steadily on his way, convinced that the way his military instincts marked out for him was the one sure way to victory, and that, if we did not flinch, victory in the West would still be possible in 1918.

I shall always regard the winter of 1917-18 as the period of Haig's greatest anxiety. He did not readily reveal his troubles to others; but occasionally in conversation he gave me hints of some of them, and I learned much from talks with members of his personal staff. It was clear that Germany was free if she cared to launch a huge offensive in the spring, and that the fighting might well be decisive. Meanwhile the British line was being seriously weakened—the strength of infantry divisions was reduced from twelve battalions to nine, five divisions were withdrawn to meet a German threat in North Italy, and under French pressure the British army was required to take over an additional twenty mile front without adequate provision being made to man it. Haig had other anxieties of a more personal kind. Mr Lloyd George did not conceal his distrust in the military direction of the war; he even went so far as to give public expression to it. Fully alive to the disastrous effects that divided counsels and lack of confidence were certain to breed, Haig let it be known in authoritative quarters that, if the Prime Minister did not wish him to continue as Commander-in-Chief, the change ought to be made with the least possible delay. The Prime Minister did not dare as yet to take so drastic a step, but he found other ways of limiting Haig's authority. With the laudable object of coordinating Allied strategy in the various fields he had already secured the establishment at Versailles of a Supreme War Council. Haig for his part was willing to believe that such a central authority might serve certain useful purposes; but he was convinced that it was not the sort of body to act effectively in an emergency, and that it might prove a serious danger if invested with executive powers. As I heard him express it many a time: 'I can deal with a man, but not with a Committee.' Then in February Sir William Robertson was adroitly removed from the post of CIGS, and his place as chief military adviser to the Government was given to the ingratiating and accommodating Sir Henry

Wilson, whose military judgment was not rated highly among his army colleagues.

Men in responsible positions must be prepared, if need be, to stand alone; and Haig was a lonelier man than most. A weaker man would have become a prey to doubts and uncertainties. But there was no moodiness or vacillation about Haig. Like King Alfred in Chesterton's poem,[1]

> 'He saw wheels break and work run back
>   And all things as they were;
> And his heart was orbed like victory
>   And simple as despair.'

Popular imagination pictures a Commander-in-Chief as giving his whole time and thought to problems of military strategy and to constant meetings with those to be engaged in the fighting. By contrast Haig was obliged to devote no small part of these opening months in 1918 to prolonged conferences and consultations in Versailles and London.

In this connection it is worth recalling that he was again in London from March 13th to 16th, on the eve of the great enemy attack, and attending a meeting of the Supreme War Council at which (among others) Lloyd George, Clemenceau and Foch were present. Meanwhile there was taking place an event of a wholly different category which was to have a profound significance for him in his personal life. In the midst of all his preoccupations and anxieties word was brought to him of the birth on the 15th of a son and heir; and on the following morning he had just enough time to see his infant son before hurrying back to France to await the coming storm.

The storm broke on the morning of Thursday, March 21st. Dense fog in the battle zone reduced visibility to a few yards. After an intense bombardment the Germans attacked in overwhelming strength, smashed the defence of General Gough's Fifth Army, and despite tough resistance continued day after day to drive furiously ahead. Viewed as a defeat, this was the biggest defeat in British military history. But it was a defeat that had in it the seeds of victory.

[1] *The Ballad of the White Horse*, Book vii.

On the Saturday evening I went round to the General Staff Office at Montreuil, hoping perhaps to get some light on how the situation was developing. All that I learned made it plain that it was becoming more and more critical; and when Sunday came I felt it altogether unlikely that I should see the Chief at church. As it happened, all the services that morning (I had always two additional ones in near-by camps) were to be taken by another minister of the Church of Scotland, the Rev. Dr James Black of Edinburgh; so it was with a comparatively free mind that I waited outside the Church hut to greet the Chief if he should arrive. I could scarcely believe my eyes when his car appeared as usual. Before the service began I had a short talk with him of peculiar intimacy; but that is a story which I reserve for a later chapter.[1] And in the course of a more casual talk when the service was over he suggested (much to my surprise, I may say, in view of all that he had on his mind that day) that I should come to lunch.

The lunch party at Beaurepaire Chateau on that Black Sunday was a small one. The Chief had as usual an ADC with him; the only other guest was Brigadier General Cox, the brilliant young General Staff officer who was reputed to be working like a Trojan in his new post as Chief of the Intelligence Department, and who, when at last a breathing space came in July, was to meet a tragic death while bathing. One of the minor surprises of the enemy attack was that a few shells were now being flung each day into Paris; and in view of the many wild theories that were circulating about this, it was interesting to me to hear authoritively about a new piece of artillery (Big Bertha) which the enemy had set up with a range of seventy miles.

Lunch over, Haig and Cox retired, and were not to be disturbed; and I had a talk with the ADC until Cox should be free to take me back with him to Montreuil. As we waited, the telephone rang; it was from the Supreme War Council at Versailles, and the reply was that for the present the Commander-in-Chief was engaged. A few minutes later the ADC was summoned to the Chief's room, and returned with an

[1] p. 120.

order to the Fifth Army authorizing a further withdrawal as far as Longueval.

The Chief had had little to say at lunch; he clearly had something very serious to occupy his thoughts. But I was impressed, as always, by his calm and resolute bearing; and as always something of his calmness and resolution spread to others. Mr Lloyd George in his *War Memoirs* has depicted 'the depth of dejection' in which, as he says, Haig and Pétain were both floundering until 'the supreme courage of Foch saved the situation'. This *ex parte* statement is completely false in its picture of Haig. There was no 'floundering in dejection' on his part. But a new factor of the utmost gravity had now emerged. Taking stock of this, Haig recognized that it could only be met by urgent and drastic action. And, for his own part, he already saw clearly what would have to be done, and was prepared to do it.

The new factor was the doubtful attitude of our French ally. When Haig and Pétain had objected strongly to handing over control of reserves to the Supreme War Council in Versailles, there was a clear understanding between them that each should come to the other's help if either army was unduly threatened. But Pétain had seemed very half-hearted in his support when Haig had gone to see him on the Saturday afternoon. He professed readiness to try to maintain contact with the British right flank; but when Haig urged that a strong French force should join immediately in blocking the road to Amiens, he declined, pleading that he expected an early attack to be launched against the French in Champagne. The truth is that Pétain had been completely misled by the enemy's stratagems, so that his 3rd Army—theoretically, as promised to Haig, supporting the British right flank—was actually dispersed by him along the French front to defend it. To the British Commander-in-Chief this fear on Pétain's part seemed wholly unrealistic—the number of divisions engaged showed that the Germans were throwing their full weight into the thrust towards Amiens, and had neither men nor guns to spare for a massed attack elsewhere. Pétain's attitude filled Haig with acute misgiving. Something very sinister seemed to lie behind it. Had the French Commander come to accept the round-up of the British forces as inevitable, so that his immediate concern now was the safety of his own army and the defence of the French capital? These, I am sure, were the anxieties that lay so heavily on Haig's mind that Sunday morning.

His worst fears were soon to be confirmed. He set out on Sunday afternoon to visit Army Headquarters, and went on in the late evening to put the urgent needs of the situation again before Pétain. The meeting took place at eleven p.m. in the village of Dury, south of Amiens. Pétain now made it plain that, if the British position got worse (and apparently he expected this) French troops on the British right might be withdrawn so as to cover Paris. To Haig this was a radical denial of the first principles of Allied strategy; the Allied line would be breached, and the British forces thrown back on the Channel ports. And so he hurried back to Montreuil, and got a message through at once to the War Office, urging that steps should be taken without delay to secure a commander responsible for the maintenance of the whole Allied front. He did not know that the British Prime Minister had already reached a similar decision, and that on his initiative Lord Milner had by now arrived in France (to be followed later by Sir Henry Wilson), with authority to take all necessary action.

This was for Haig a decisive hour; it was to have a parallel in 1940 when Mr Winston Churchill hastened to France in a desperate effort to preserve Franco-British unity in face of another German 'break-through'. On Monday Lord Milner and Sir Henry Wilson were fully occupied in Paris exploring possible lines of action; and conversations with Haig followed next morning. In these talks Haig removed all possible doubt about his own attitude; the first essential was to keep the Allied line intact, and so save Amiens, and with that end in view he would be content to serve under any commander who was prepared to fight. He was also persuaded that, as the new post must obviously be given to a Frenchman, the man for it was Foch, and not Pétain; he has noted in his diary how, at the Conference which followed, 'Pétain had a terrible look. He had the appearance of a Commander who was in a funk and had lost his nerve'.

At midday on Tuesday, March 26th, in the little market town of Doullens, midway between Amiens and Arras, leading representatives of France and Britain sat down round the table for one of the most fateful conferences of the war. President Poincaré was in the chair; and with him, on the French side, were Clemenceau (Prime Minister), Foch (from the Supreme War Council) and Pétain. Agreement was at once reached on Haig's primary point, that everything must be done to stop the German advance before Amiens. But on the next point, the nature and extent of French cooperation, the Conference ran into difficulties. Milner, in the Memorandum he submitted to the Cabinet, has told how Pétain continued to be discouraging. Haig, on the other hand, welcomed the proposal that Foch should be given authority to coordinate the action of the two armies, and (when this had been accepted by the meeting) went on to suggest an extension of Foch's powers to cover all allied forces on the Western front. On this cordial note the Conference broke up, well satisfied that something vital had been accomplished towards reestablishing Allied unity and blocking the German advance. Henry Wilson has recorded in his diary: 'Douglas Haig is ten years younger tonight than he was yesterday afternoon.' It remains to be added that, with Haig's cordial approval, Foch's position and powers were more clearly defined at a further conference held on April 3rd, when he was entrusted with 'the strategic direction of military operations'.

Haig's part in these fateful negotiations has not always been appreciated to the full. Sometimes indeed it has been scandalously misrepresented. It has been alleged against him that he had consistently thwarted the Prime Minister's efforts to secure unity of command; and now, when he was at his wit's end, he was forced to accept it. This is a complete travesty of the truth. Unity of command is not a magic formula, calculated without fail to promote harmony and secure success; and in the course of a long and sanguinary war, in which the manhood of two or more nations is fully engaged, a hasty agreement might easily have proved disastrous. How would either the British or the French people have reacted if their losses in the previous years had been sustained under a foreign

F

commander? But if unity was not to be easily established, the closest cooperation was from the first imperative; and for Haig it had always been a fundamental obligation to promote it. When he took over the supreme command in 1915 he made it plain that the fact that he was not strictly under General Joffre's orders made no difference, 'as my intention was to do my utmost to carry out General Joffre's wishes on strategical matters, as if they were orders'. And he faithfully adhered to that intention both with Joffre and with Joffre's successors. Serious differences of opinion no doubt arose at times, as was indeed inevitable; and Haig's practice then was to go if possible in person to see the French commander and talk things over till agreement was reached.

Mr Lloyd George's earlier efforts to secure fuller unity had failed because they were not sufficiently realistic. The Supreme War Council, whatever its uses, was not the sort of body (on this both Haig and Pétain were agreed) to be safely entrusted with the control of reserves in an emergency. The earlier Nivelle episode, by which the British Army was to be merged with the French under the existing French commander, had been a ghastly blunder in all its aspects, and ought never to have been contemplated. The situation that had now emerged was essentially different. Haig and Pétain both retained their independent commands under Foch as 'General-in-Chief'. And Haig had the right of appeal to his Government if in his opinion his army was endangered by any order of General Foch.

This was an arrangement which Haig could accept; and it is a mark of his greatness that he accepted it, not with passive acquiescence, but with whole-hearted support. He saw it as essential if disaster was to be avoided; and with it there opened again the prospect of ultimate victory. And that was all he cared for. He did not see in it any lessening of responsibility: rather, as we are to see, his responsibilities were in a measure increased, for loyalty to Foch did not absolve him from the exercise of independent judgment. And it is wholly false to represent him as having been superseded; for supersession implies failure, whereas this was a very special arrangement to ensure Allied unity. We would be nearer the truth if we saw in it a 'sacrifice' on Haig's part; only Haig was too much of a greatheart to think of it as such; he would have said he was merely doing what duty and honour demanded. His action did, however, entail sacrifice of a very real kind; for one result of it has been the popular delusion, fathered by the British Prime Minister, that the successes which followed in the next eight months were largely due to Foch. Unity of command was of vital importance so long as the Allied line was threatened; it contributed comparatively little to the subsequent advance.

The immediate gain from Foch's appointment was that Pétain was now under orders; the Allied line was to be treated as a whole, and kept intact. And the German advance to Amiens was stayed. But the British Army had now to face a new threat. Ludendorff switched his attack further north towards Arras. Here a violent enemy offensive was repulsed with fearful loss; and I recall how, on the following Sunday morning, the Chief (who rarely allowed himself such liberty of speech before the church service) told me how, thanks to the steadiness and resolution of our men, the enemy had gained little more than a few yards. His anxieties however increased with the threat of still another northern enemy offensive, directed towards the Channel ports; and there, as he knew, he could not afford to give any ground. On April 9th the enemy launched a fierce attack on a front partly held by Portuguese troops; a gap was opened which the exhausted British divisions, with no reserves available, succeeded for a time in closing. Haig's anxieties at this time were almost overwhelming. With his depleted forces (their ranks now filled by the latest recruits, ill-trained and in many cases too young and inexperienced for such an ordeal) he had at one and the same time to maintain contact with the French and block the way to the coast. The British troops had now been continuously engaged since March 21st, meeting the full force of the enemy offensive; yet Foch had persistently refused to allow French divisions to take over any part of the British line—was he afraid (as Haig was driven to conjecture) to trust them in the battle-front?

In the judgment of Mr Winston Churchill[1] this was 'prob-

[1] *The World Crisis*, II, iv. p. 433.

ably, after the Marne, the climax of the war'. He recalls how
the enemy were determined to battle the life out of the British
Army, and how the French seemed in British eyes 'sunk in
stupor and passivity'. For Haig those four days April 9th to
12th were as dark and ominous as any in the long campaign,
comparable to the recent Sunday in March when he saw his
army being left in the lurch by the French commander, or that
October day in 1915 (the memory of which still haunted him)
when a breach in our thin line seemed to have opened for the
enemy the way to Ypres and the coast. On those two earlier
occasions, as we have seen, his indomitable spirit found
expression in personal intervention; was there anything
similarly personal that he could do now?

Early in the morning of April 11th, perhaps after a restless
night, he retired to his room, and in his own handwriting
addressed a personal message to his troops. Characteristically
it was headed: 'To all ranks of the British Forces in France'—
he always preferred the inclusive 'all ranks' to the alternative
'officers and men'. In the simple straight language which came
naturally to him on paper he explained in a few short paragraphs how critical the situation had become. Words failed him,
he said, to express his admiration for the determination and
self-sacrifice of the British troops during the fierce fighting of
those last three weeks. Victory would now belong to the side
which held out the longest. With pardonable exaggeration he
added, in the interests of encouragement: 'The French Army
is moving rapidly, and in great force, to our support.' And he
closed with the stark reminder that we stood with our backs to
the wall, and that each one of us (that individualizing phrase
occurs twice in the closing sentences) must fight on to the end.

I have been interested to find that in his diary for that day
Haig makes no reference to the writing of this Order; there is
little indeed to indicate the depth of his anxiety at that
particular moment. Some measure of his apprehension, however, may be gauged from the words: 'We must expect the
enemy to press his attacks to the Calais-Boulogne coast; and I
am still anxious to have French support about Cassel, since
Foch refuses stoutly to take over any line from us and to set
free British divisions.'

As if by a miracle the situation was saved; Ludendorff, within an ace of success, was foiled again by the refusal of our men to yield. A week later four French divisions took a place at last in the British frontline, and were assigned the defence of Mount Kemmel; and it was a deep disappointment to Haig that this commanding position, which his own depleted and exhausted divisions had held for so long, was lost by the French at the first fierce assault. Ludendorff, however, failed again to exploit his success; and with this failure the threat to the Channel ports came to an end.

Tensions and disappointments, however serious, were never allowed by Haig to cloud unduly his outlook or to poison personal relationships. But during this period his equanimity was tried almost to breaking-point. In May Foch borrowed five badly-battered British divisions to hold a quiet part of the French line, and when he stationed them on the Chemin des Dames he brushed aside Haig's strong objections that the enemy were known to have made preparations for an attack on that front. The threatened German attack was duly launched on May 27th with a ferocity and a success that almost threw open the way to Paris, and in the course of it four of the five British divisions were overwhelmed and their total strength reduced to that of one. Serious criticism of both Foch and Pétain began at this time to manifest itself in French government circles. The British Government too became deeply disturbed about Foch's disposal of British reserves; and when, in mid-July, Foch asked the loan of four British divisions to help in countering a serious German attack against the French on the Marne, they practically invited Haig to decline.

This was a delicate position for the British Commander to be placed in. He knew, as the Government also knew, that the enemy was ready at any moment to launch a new attack on the British front; and with a Prime Minister still disposed to be critical a false decision on his part was likely to recoil on his own head. A lesser man would have chosen the easy way out. Not so Haig; his reading of the military situation convinced him that the position on the French front was the more urgent. And so he reached a gentleman's agreement with Foch to meet the immediate need, and the divisions were sent.

This was a decision characteristic of Haig at his best. And it was a decision that justified itself; for by a strong counter-attack in which the conduct of the British troops came in for the highest praise, the French succeeded in driving the enemy back to the Aisne. This marked at last the turning of the tide. The initiative was slowly passing into the hands of the Allies.

Those last four months March to July, and more especially the months March and April, will always stand out as among the most critical in the history of the British Army. Relentless and unceasing attacks left us constantly poised on the brink of defeat. If defeat was averted this was a triumph made possible by the courage, tenacity and endurance of 'all ranks'; and among the Allied commanders it was in a very special measure a triumph for Haig. It is a tragic reflection that during this agonizing period there was little evidence that he and his army colleagues in France had the confidence of the British Prime Minister. As Mr Lloyd George interpreted the situation, it was they who were responsible for the defeats; and right on to the end of July (as Lord Hankey has recently made plain) he was still searching for some way of displacing Haig. Fortunately there were wiser and steadier heads in the Government. One was Milner, now Secretary of State for War, whose apprecia-tion of Haig grew steadily with closer acquaintance. Another was Mr Winston Churchill. Both of these were frequent visitors to GHQ in those anxious months; and I recall in particular one Sunday evening in April when I sat next to Mr Churchill at dinner as he sat next to Haig. In my ignorance I had been inclined to view Mr Churchill's presence on that occasion with some apprehension—Gough had been recalled to England, despite a strong protest from Haig; was Haig perhaps to be the next victim? My apprehensions with regard to the guest of the evening were entirely baseless. Since the opening of the March offensive Mr Churchill, who was Minister of Munitions, had been insistent in seeing that the army in France had all the munitions, as well as all the men, it required; and Haig, as he records again and again in his Diary, was very appreciative of his efforts. And Mr Churchill has himself given us a picture of his own more personal feelings at one of the darkest hours of crisis. The time was during that week in April when our backs were to the wall—had he perhaps just seen or read Haig's Order of the Day? And this is what he writes[1] : 'For all the criticisms I had made and all the convictions I held about the Somme and Passchendaele, my heart went out to the Commander-in-Chief as he bore the trial with superb and invincible determination. On the night of the 12th I telegraphed: "I cannot resist sending you a message of sympathy and sincere admiration for the magnificent defence which you are making day after day, and of personal confidence in the result".' Here deep spoke to deep. And the man to whom those heartening and appreciative words were addressed had been at his post, bearing with unconquerable soul an unprecedented load of responsibility, from the first week of the war until now; and he was to be there till the end. We may well apply to him those lines of Robert Browning :

'Here had been, mark, the General-in-Chief,
Thro' a whole campaign of the world's life and death,
Doing the king's work all the dim day long.'

[1] *The World Crisis*, II, iv, p. 438.

CHAPTER 10

1918 (ii): ON TO VICTORY

The worst part of the long fight was now over. The time had come to reap the harvest of the months and years of agony that had gone before. The victories which were now to follow in quick succession were in the main British victories, fought by Britain's citizen-troops and directed by British commanders. And, in accordance with the aim which Haig had steadily pursued and the hope which had consistently inspired him, the war was to be brought to a dramatic close before the year was out.

Having by July fought the enemy to a stand-still, what was now to be the policy of the Allies? Foch's mind was set on clearing the enemy from the Bruay coal-fields and other important strategic areas as a necessary preliminary to a big offensive in the summer of 1919. The policy of the British Government was similar; it may be gathered from a lengthy document[1] prepared in July by their chief military adviser, Sir Henry Wilson, which envisaged limited Allied attacks in 1918, a possible German attack in Italy in the spring, and an all-out Allied offensive in July 1919.

The one man in a responsible position who had a completely different outlook and policy was the British Commander-in-Chief. For over a year he had hoped and planned for a victory not later than the end of 1918; and despite the shattering setbacks of 1917 and the subsequent enemy offensives he had never allowed that hope to die. Even before the fierce German offensive of March his strong conviction, expressed on several occasions, was that if the enemy should attempt a breakthrough and fail, their failure would mark the beginning of the end. Now their failure was apparent even to themselves.

---

[1] Haig's copy of this document may be found among his private papers. Scribbled over it in his hand-writing is the contemptuous comment: 'Words! Words! Words! lots of words and little else. Theoretical rubbish! Whoever drafted this stuff would never win a campaign.'

Haig's policy was to strike hard, and to give a demoralized enemy no time to recover.

With some reluctance (for his own ideas had been different) Foch acquiesced in Haig's selection of a *locale* for the first attack, to the east and southeast of Amiens; and he agreed to place the First French Army under Haig's command. This shock attack on August 8th, planned and carried out with the utmost secrecy and care, was one of the most brilliant successes of the war. Ludendorff in his *Memoirs* describes August 8th as Germany's 'black day', and he attributes to this defeat the general discouragement that set in among Germany's allies, and the submission that followed on September 30th of the first of them, Bulgaria. Lord Hankey tells how the victory had another significant result. The War Cabinet no longer entertained the thought of a possible change in the British High Command; and he goes on to say that he was glad of this, 'as I never discovered any other officer of the same calibre as Haig'.

After a few days Haig decided to break off the attack and to switch it further north; his policy was to shatter the enemy's resistance by a series of onslaughts, short, unexpected, and concentric. This decision involved him for the next five days in violent disagreement with the Generalissimo—disagreement which at last was amicably settled when he went to see Foch personally, and the two commanders had a frank man-to-man talk. To quote here from his Diary: 'I spoke to Foch quite straightly, and let him understand that *I was responsible to my Government and fellow citizens for the handling of the British forces.* Foch's attitude at once changed; and he said all he wanted was early information of my intentions so that he might coordinate the operations of the other Armies, and that he now thought that I was quite correct in my decision not to attack the enemy in his prepared position.' The new British attacks were duly launched as Haig had planned, and were extraordinarily successful; within four weeks the enemy was driven back approximately twenty miles on a wide front, with the loss of 50,000 prisoners and 500 guns. Foch was generous in his praise of these successes, and of the way they had been gained; in a personal message to Haig he went so far as to say that those operations of the British Army 'would serve as a model for all time'.

Despite these remarkable victories there was little disposition on the part of Cabinet ministers or ordinary citizens to imagine that the war could be ended before another year. The enemy had now retired behind the formidable barrier of the Hindenburg Line—not a single line, as the name might seem to indicate, but an elaborate system of defences with switch lines linking up strong fortified outpost zones. It was reputed to be well-nigh impregnable; and the enemy felt safe for the winter. But a common saying of Haig's at this time was: 'Fortifications don't mean much when men have no longer the will to defend them'; and, convinced that the German soldier had lost his zest for fight, he prepared to attack.

The responsibility that lay on Haig was enormous. It would have been less onerous if he could have counted on the confidence and support of the British Government. But in all the sweeping victories of those recent weeks not a word of congratulation had reached him from that quarter. Then on August 31st he received from Sir Henry Wilson as CIGS a telegram which deserves to be quoted in full. 'Just a word of caution in regard to incurring heavy losses in attacks on Hindenburg Line as opposed to losses when driving the enemy back to that line. I do not mean to say that you have incurred such losses; but I know the War Cabinet would become anxious if we received heavy punishment in attacking the Hindenburg Line without success.'

Ostensibly the telegram was a personal message from the CIGS; and Mr Lloyd George subsequently disclaimed any share of responsibility for it. But such a message could not have been sent without some kind of authorization. One can appreciate the acute dilemma in which the War Cabinet found itself. It was haunted by memories of Passchendaele; the man-power situation had become acute; men of fifty were now being called to the colours; losses might be incurred which could not be made good before next year's decisive offensive. But a decision had to be taken now one way or the other; and apparently the War Cabinet did not wish to have the responsibility of taking it. 'In the most invidious manner' (so Mr Churchill describes the incident in *Great Contemporaries*), they cast the direct responsibility upon the Commander in the Field. Writing that night in his diary Haig refused to believe that the CIGS in his telegram was expressing a merely personal opinion. 'I read it to mean,' so he wrote, 'that I can attack the Hindenburg Line if I think it right to do so. . . . If we fail, or our losses are excessive, I can hope for no mercy!'

This was the hour when Haig rose to his greatest height. There was no one in authority, military or civilian, with whom he could share his responsibility. He stood absolutely alone. Had he shirked the issue, no blame would have come to him —except from his own conscience. Without hesitation, he accepted the challenge. He was cheered by a visit from Mr Churchill on September 8th; and next day he crossed to London where, at his own request, he saw Lord Milner at the War Office, and laid the whole situation before him with prophetic insight and conviction. The whole character of the war, he said, had changed; and 'if we act with energy now, a decision can be obtained in the very near future'. There was no resisting such an appeal. The War Cabinet acquiesced in the proposed attack, though according to Sir Henry Wilson Milner continued to think Haig 'ridiculously optimistic'.

During September and October French, American and Belgian armies all joined in an offensive that spread from the Channel to Lorraine, and all shared in the sacrifices and the successes of the final battles. But everything at this point hinged on the British attack on the Hindenburg Line. Here the decisive battle opened on September 27th; the formidable Canal du Nord, with its banks sometimes sixty feet deep, was successfully stormed; further south the Canal de Saint-Quentin was crossed with the aid of collapsible boats, rafts and life-belts; and, forced from their last strongholds, the Germans began a withdrawal along the whole front.

The results of this outstanding victory were swift and dramatic. On October 3rd the German Government appealed for an armistice—an appeal which Foch, in reporting it to Haig, generously acknowledged to have followed directly from the British piercing of the Hindenburg Line. Meanwhile one satellite power after another, Bulgaria, Turkey and Austria, read the writing on the wall, and concluded an armistice. Discussions with Germany went on for some weeks; and an armistice was finally signed on November 11th. The war was ended.

During these closing months of the campaign the British Commander-in-Chief (attended only by three or four members of his personal staff) had been living in a railway train in the forward areas—moving from place to place as the situation developed, visiting units engaged in the fighting, keeping in close personal touch with his Army Commanders, and exercising all the time direct control over operations.

I had been absent during September on sick-leave; and after my return to Montreuil I went forward on October 10th to call on the Chief. In his absence I had a long talk with Colonel Alan Fletcher, who was frank and communicative as usual. The Chief had gone to see Foch. He was most eager, I was told, to have more troops (perhaps American, if that could be arranged) to help in a strong forward push beyond Le Cateau; our own men had now been fighting so long and so hard without rest that he did not feel he could put a new and heavy demand on them. If only he had the men, he was willing to pledge his military reputation that we could be in Brussels in a fortnight.

My next opportunity to see the Commander-in-Chief was on the morning of November 13th, two days after the Armistice was signed; his train was then in a siding near Iwuy, a few miles northeast of Cambrai. When I arrived he had been talking to a Pipe-Major of the Gordon Highlanders who had come over from a near-by Scottish Division to enliven the morning air with the music of the pipes. He turned at once to welcome me; and when I attempted to express congratulations his immediate reaction was: 'Oh! you mustn't congratulate *me*; we have all been in this together, all trying in our different ways to do our part.' After a time he added, pointing to the piper: 'Come and speak to this fine fellow over here. He came out in 1914; and in the early days was through some of the worst of it. It is fellows like him who deserve congratulations.' A trifling episode in itself, no doubt, but also a revealing one. Haig was a genuinely humble man; and in the hour of victory he identified himself with the countless unknown men and women who by their heroism and endurance had done their part to purchase victory.

After this he took me into his little working-room in the train. As we sat there I noted how fresh he looked, thinner perhaps and less broad shouldered than he once was, yet still lithe and vigorous; and I reflected how well he had stood up to the strain of those terrible years. There was something in Haig which remained unchanged, in storm or in sunshine—a quiet dignity and poise, and along with it a deep humanity, which the outsider was not always privileged to see, but which readily came to the surface on an occasion such as this. We talked of home—of Lady Haig, on whom I had recently called at her residence in Kingston Hill, and the infant son, and the two little daughters, who were so pleased that the end of the war had given them a holiday at home from school. I asked him if he had any thought of following the enemy into Germany. 'None at all,' he replied: 'we have beaten them in fair fight, and that is enough for me.' General Plumer, he said, would be going up with an Army of Occupation, and any official entry could be left to him. There was no exaltation, no bitterness, and above all no trace of hate, as he spoke of the enemy's overthrow. 'Poor wretches' was a phrase he used, as he recalled both the great fight they had put up and the demoralization that had ensued. But he was glad that the fight was over; if the end had not come now they might have decided to put up a last despairing resistance at the Meuse in spring; and that would have been costly to all, more especially to the British, who would have had to do most of the effective fighting. As to the terms of the armistice, which had largely been dictated by the French, he commented, 'Personally, I would have let them off more easily.'

We then talked about a Church service on Sunday. He was likely to be on his train for another week, for there was much to be attended to. But as the 51st (Highland) Division were stationed not far away, he would be happy if he could go to worship with them. So he invited me to be his guest for the weekend, adding that we could then go to the service together. I am afraid that few who were at the open-air service on the

Sunday are likely to have retained pleasant memories of it. It was a bitterly cold November morning. The divisional authorities had arranged a very large parade; the men had been in their place unnecessarily early; and it was a relief when the time came to disperse.

In the train, where I was the only guest, the Chief was busy, for most of the day, acknowledging hundreds of congratulatory messages. He told us at dinner of a telegram from the Portuguese War Minister who had expressed his praise for 'Britain's glorious troops'. 'I didn't feel in my reply,' the Chief said with a smile, 'that I could rise to "glorious" about his men, but I managed "gallant".' Incidentally I may add that I never heard Haig speak disrespectfully about the Portuguese 'rank and file'. They might have fought very well, he said, if they had been under better officers. One subject on which he spoke to me with great seriousness was the provision to be made at home to welcome men on demobilization. He pictured them arriving in hundreds at some of the big railway stations, with perhaps nowhere to spend the night. 'That is a matter,' he said, 'in which the Churches can help a great deal, and I hope they will start to make provision at once.' Then he added, without going into the matter further: 'Of course that is only part of a very much bigger problem.' It was significant that even at that date he was already giving thought to the post-war needs of the men he had commanded.

The Chief and his personal staff were to leave the train on Tuesday to return to Montreuil. When that last morning came, he invited me to accompany him for a short after-breakfast stroll across the fields. My thoughts, I recall, centred on what it must mean to him, as in fact it meant to us all, that the ordeal of the last four years had now ended, and ended so triumphantly. But it was fundamental to all Haig's thinking on the war that the present was integrally bound up with the past, and that victory would have been impossible apart from the sufferings and losses which had preceded it. And so in reply to something I said he recalled the terrible experiences of the men in the trenches—how little they could see at any time beyond the day-by-day happenings, and how extraordinarily hard it must have been amid all the set-backs and

losses to maintain a confident hope in final victory. Then he
told me about his own outlook. With something of that quiet
confidence which so often characterized his deepest moods he
said quite simply: 'I always saw that it would come to this.
I never doubted the final outcome.' And then, as if in corrobora-
tion of his faith, he added, after a pause, the significant military
comment: 'That is why I kept some of my cavalry.'[1]

Our small company, the Chief, two members of his personal
staff, and myself, set off a few hours later in two cars. We
passed through Cambrai; and near Marquion on the way to
Arras we left the cars for a time, and I accompanied him as he
surveyed at close quarters scenes of much fierce fighting in the
September advance. Before we reached Arras a dense fog
descended; and from there to Montreuil the cars proceeded at
a snail's pace. Late that evening, quietly and unobtrusively, the
Chief came 'home' to the Chateau de Beaurepaire.

[1] Haig's faith in the possible use of cavalry has exposed him to much
criticism. But in the closing months of the war, when the enemy
were badly demoralized, cavalry proved an effective asset.

CHAPTER 11

THE POST-WAR YEARS

Mr Duff Cooper finely described Haig's life as an epic drama of four years and one hundred days, with a preparatory prologue of fifty-three years and an epilogue of ten.

During those post-war years I saw him often, and had many letters from him; but I propose to say little about that here. His work in France did not end with the signing of the Armistice. He had still many official duties to perform, and many responsibilities to discharge towards the Army he had commanded. Among subjects that were much on his mind were demobilization and pension allowances. And he sometimes spoke to me of the dangers ahead if the peace settlement should be made one of vengeance. He was gravely distressed when the Government, setting aside completely certain well-considered views on demobilization which he had laid before them as early as 1917, adopted a scheme which would give priority to certain 'pivotal' individuals and classes, a scheme which caused intense resentment throughout the Army when it became known. And in talks with me he was loud in his praise of the energy and wise judgment with which Mr Churchill, on becoming Minister of War, tackled the whole problem afresh, the new scheme being based (as Mr Churchill has emphatically acknowledged) on the very principles which Haig had previously laid down as being of fundamental importance.

His command in France ended in April, 1919. For a year thereafter he was in command of the Home Forces. Then he passed into private life. Was he disappointed that there was now no important Government post, at home or in the Empire, where he could be of service? I do not think so. The one big peacetime task which he saw awaiting him was one arising directly from his wartime responsibilities; it was the care of ex-service men and their dependants.

For his services in the war Haig received an earldom, with a grant of £100,000. Sir John French, on his recall in 1915, had been made a Viscount; and a higher distinction was obviously due to the leader who had been at his post during the whole campaign and had brought us at last to victory. In a letter I had from him at the time he wrote: 'I am proposing to take my title from the homeland—a kind of compliment, I hope folk will think, to what our brother Scots have done in the Great War. This seems to me far better than to describe myself as of some small spot in the far-flung battle line of our great armies in France and Flanders.' Bemersyde was the ancestral home of the Haig family; 'Tide what may, whate'er betide, Haig shall be Haig of Bemersyde.' He was therefore deeply moved when a spontaneous movement on the part of friends at home and throughout the Empire led to his being presented with the ancient Border 'keep' of Bemersyde; and this, extended and reconditioned, was now to be his home for the rest of his life. He had seen little of Scotland since boyhood; but it was often remarked among those who knew him well that in those post-war years he reverted more and more to the traditional ways of life and thought of the Scottish people.

On his return from France honours of all kinds came to him from a grateful people, so that for the next year life became for him a triumphal procession of a kind which he would not readily have chosen for himself. He welcomed it, however, for the pleasure it gave him to meet so many of the men who had served under him, and as an opportunity to remind public authorities of the duty they owed to ex-servicemen, and of the needs to carry forward into peace the lessons learned in war.

An engagement which gave him peculiar pleasure was to be installed as Rector of the University of St Andrews. He had known St Andrews well as a boy, and had been for a time at school there. In the Scottish Universities the Rectorship, originally a mediaeval institution, is now essentially a post of honour, held for three years by some person of distinction whom the students select to be their official representative. I recall the undisguised elation with which, one Sunday morning in 1916, Haig said to me: 'I have an invitation from the students of St Andrews to be their Rector; I never had such an honour done me in my life.' The invitation had come to him during the long-drawn-out Somme battle; and that at such an

G

hour a body of youth in the homeland should spontaneously offer him so signal a token of their appreciation was like a breeze from the hills blowing fresh about his spirit.

The Rector is expected to deliver an Address on the day of his installation; and some of the Rectorial Addresses at Scottish Universities have been memorable pronouncements. This was something of an ordeal for Haig; but he responded to the welcome of his student-audience and soon found himself very much at ease with them. Basing his address on the famous funeral speech of Pericles at the time of the Peloponnesian war, he went on to examine the qualities which had brought our nation successfully through the greatest ordeal in its history, and which would be no less necessary in the years ahead. It was a characteristic utterance, with not the least trace of egotism; from first to last it was a tribute to the valour of the British soldier, and a confession of faith in the character and destiny of the British people. Three years later the Chancellor-ship of the University of St Andrews became vacant; and Haig was the obvious choice for this higher honour. On the same day as J. M. Barrie, the new Rector, delivered to the students his famous address on Courage, Haig was installed as Chancellor.

But punctiliously faithful as Haig was in any duty that he undertook, his first care now was for the men and women who had given their all in the service of King and country. Here we come to a story that will live in history alongside that of his command in war. And it ought to silence for ever the suggestion that Haig was lacking in human sympathy and under-standing.

Haig never imagined that his responsibility towards the men who had fought and suffered with him ceased with the end of war. He had a vivid perception of the problems that were to arise for many of them, more especially for the disabled, on their return to civil life; and ever present to his mind was the shadow of fatherless homes, of unhealed wounds, of financial insecurity and the threat of unemployment. Profoundly conscious that State aid, however generous, was not in itself sufficient to meet the need, he never ceased to lay the burden fairly and squarely on local authorities, on the churches and other public bodies, and above all on the conscience of the individual citizen. The whole country responded to his lead. In 1921 the first Poppy Day appeal was launched. It had been somewhat hurriedly arranged; and Haig privately gave his own guarantee against possible loss. Before long Poppy Day developed into a vast co-operative and voluntary effort which has brought in many millions of pounds to relieve the sufferings occasioned by war. In all this the nation awoke to a new image of Haig; the victorious commander was now seen to be preeminently the soldier's friend.

But the greatest monument to Haig's peace-time labours was the founding of the British Legion. The war had scarcely ended before a number of ex-servicemen's associations came into being, with conflicting aims and in some cases with a strong political bias. Haig saw in this a disastrous situation, opposed to the best interests of the men themselves, and certain to occasion national unrest and strife. And so he set himself to deal with it. Ill equipped as he was for the task of swaying an audience, it is a striking tribute to the esteem in which he was held, as well as to the sympathy and sincerity of his appeal, that he succeeded in persuading those various associations to sink their differences and to form one united society, the British Legion, in which men who had served their country in war should now, without regard to service rank or political affiliations, work together to further the aims which as ex-servicemen they held most dear. It is safe to say that no one else but Haig could have done it. From the first he inspired the new organization with his own ideals of unity, comradeship, service, and the promotion of world peace, and he imparted to it a character which has never left it. Today, after a second world war, the British Legion still carries on its beneficent work in accordance with the spirit of its founder and first President.

Encouraged by his success in the homeland Haig turned his attention next to the Dominions. Difficulties there were more formidable; but by the sheer weight of his personal influence he brought about a similar unification of societies in South Africa and in Canada, and laid the foundation of the British Empire Service League. His hope of being able to visit

Australia and New Zealand was not to be fulfilled. Meanwhile he was untiring in the work that fell to him at home. He had a vast correspondence, to most of which he preferred to reply in his own handwriting; for with all his reserve Haig never lost the personal touch. Those of us who saw him from time to time sometimes feared that he was overtaxing his strength. On the night of Sunday, January 29, 1928, when he was on a visit to London, he was suddenly seized with a heart attack; and in a few minutes all was over.

It is hard to convey to a later generation the emotion that swept the country when the news went out that Haig was dead. Not within living memory had the nation accorded to any of its sons such a demonstration of loyalty, gratitude and affection. Day after day thousands filed reverently past the body as it lay in state, first in London and then in Edinburgh. Vast crowds lined the London streets as the funeral procession went on its way from St Columba's (Church of Scotland) in Kensington to the national service in Westminster Abbey, and then to Waterloo Station, from which the body was to go by train to Edinburgh. It arrived there at midnight; snow had fallen; and as the gun-carriage on which it rested proceeded on its silent way, along dark streets and under the brow of the Castle Rock, to St Giles' Cathedral in the High Street, the route was lined by denser crowds than had ever been seen even for a royal visit. Not since the burial of the Regent Moray in the sixteenth century (so writes Dr Charles Warr, Minister of St Giles' and Dean of the Thistle) had Edinburgh witnessed scenes comparable to those that were enacted on this occasion. Some days later the body was taken by train to the Scottish Border country which Haig had come to love so well; then, resting on a farm cart drawn by four farm horses, and accompanied by a solemn concourse of folk from the immediate neighbourhood and by representative members of the British Legion, it passed without any semblance of military pomp to what was, in accordance with Haig's own wish, to be its last resting place, quite near to Bemersyde, in the grounds of the ruined Abbey of Dryburgh.

Those demonstrations on the part of the British people were all the more significant because Haig had never sought to project his image on the popular imagination. It would not have been surprising if a man so reserved had gone to his grave, honoured indeed for his public services but unaccompanied by any widespread manifestation of grief or affection. In the nation's tribute to Haig there was without question a grateful recognition of his war-time leadership, but even more there was a deep reverence for the man himself. What sort of man could this be (so people had been quietly asking themselves) who, coming home to the honours and the leisure which were recognized as his due, could find complete satisfaction in neither, but must wear himself out in the service of the men he had once commanded? They noted, too, that though he was prepared at all times to lift up his voice on behalf of others, he had never a word to say in vindication of himself. And every story that they heard of him from time to time, however homely and trivial the setting, confirmed them in the belief that this man was made in no ordinary mould; he had standards of character and conduct which lifted him high above the common level.

It is not amiss to recall how men felt about Haig in his death; it has much to tell us of the honour in which they held him in life.

III : PERSONAL

CHAPTER 12

THE MAN WE KNEW

I return to the question : what sort of man was Douglas Haig? The best answer may be found in the tributes paid to him, during his life and at the time of his death, by men who served with him and knew him intimately. I recall more particularly the Memorial Number of the magazine *British Legion* for March 1928—as impressive a collection of tributes as has appeared in praise of any great national figure in the present century.

In an earlier chapter I wrote of Haig's resolute sense of purpose; it had been a marked feature of his life ever since he entered Sandhurst, and goes far to explain his subsequent rise to preeminence, when so often he seemed to be the one man obviously marked out to undertake some new and challenging military responsibility. By contrast I turn now to look at that earlier picture which we have of him, the undistinguished schoolboy, the carefree Oxford undergraduate, and ask what can have happened to him in early youth to give his life a new direction and purpose.

There is some indication that while at school at Clifton he had entertained the thought of the army as a possible career; but the years passed, and it may have been in the hope of still securing entry through the University that, at the age of nineteen, he matriculated at Oxford. But the choice of profession does not in itself explain the purposefulness with which he now set himself to prepare for it, and which in fact from now on became an integral part of his character. Did the death of his mother perhaps serve in some subtle way to stir him into life? A lady of refined tastes and sincere personal piety, she had watched with rather special care over the development of the youngest member of her large family; and he in turn was deeply devoted to her. She died when he was eighteen (her husband had died in the previous year), and he felt her loss profoundly. Her place in his life was taken by his youngest sister Henrietta (Mrs Jameson), who was his senior by ten years, and who, at this difficult stage in his career, became for him companion, encourager and confidante, and continued to be, until his marriage, the main influence in his life. We probably owe it to her, not merely that he now decided finally to make the army his career, but that he also set himself resolutely to develop the powers which she saw to be latent in him. Did her influence perhaps go deeper still? I know of no positive evidence that the claims of religion played any part in the change which occurred in Haig's life about this time; and in the years that followed he did not seem to be more than conventionally religious. But the religious training which the Haig children had received from their mother, and from other teachers, was directed towards encouraging them to take a serious view of the purpose of life; and it would be more easy to explain the depth and sincerity of Haig's religion during his later years if it was a recapture of a religious experience (not revealed to others, and then sinking perhaps into the subconscious) which helped him at this early formative period to associate his life with some (as yet) undefined purpose, to see his course ahead, and to follow it out with resolution. Haig's sense of purpose had always in it an element of dedication.

I had many talks with him about his time at Oxford. Recalling that comparatively few of our army leaders had studied at a university, I once asked him whether as a soldier he had gained much from his experience as an undergraduate. 'Certainly,' was his reply; 'by the time I went to Sandhurst I had learned to think for myself; I didn't, like so many of the other fellows, take everything I was taught for gospel.' And he added that that was a lesson which had remained with him all his life. How well he had learned it has been abundantly attested by colleagues, military and civilian, who had come to attach the highest value to his powers of clear thinking and sound judgment. When Lord Haldane addressed the House of Lords on the occasion of Haig's death, he chose to emphasize, in sentence after sentence, Haig's supreme gifts as 'a military thinker'. He told how, when faced as Secretary of State for War with the task of army reorganization, he decided that the right thing was to ask Haig, who was then in India, 'to come over to this country and think for us. From all I could discover even then he seemed to be the most highly equipped thinker in the British Army'. Similarly in January 1917, when Haig was appointed Field-Marshal, Haldane wrote in a letter of congratulation: 'You are almost the only military leader we possess with the power of thinking, which the enemy possesses in a highly developed form. The necessity of a highly trained mind, and of the intellectual equipment which it carries, is at last recognized among our people. . . . If I had had my way, you would have taken the place at the head of a real great Headquarters Staff in London on August 4, 1914.' In face of such tributes we may dismiss as ludicrous the myth which was sedulously spread abroad by Mr Lloyd George that Haig was essentially 'stupid'.

One of Haig's tutors at Oxford had been Mr Walter Pater, the distinguished literary critic and stylist. I had long been familiar with Pater's *Marius the Epicurean*, and I confess I saw something incongruous in the fact that the destined leader of our nation's armed forces in this greatest of wars should have received part of his training at the feet of this fastidious aesthete and master of English prose. I once ventured to ask whether he recalled any influence that he had derived from Pater. 'Oh!', he replied, with a smile of self-depreciation, 'Pater would discourse to me about Plato when my own desire many a time was to be out hunting.' Then, becoming more serious, he added: 'But there was one thing I did learn from Pater, and I have never forgotten it. He used to impress on us that, if we were to express our ideas fully and clearly in writing, we would need first of all to think out clearly what it was we wished to say; then we would need to be equally careful to find the right words in which to say it, for it might well be that there was only one word, or one form of words, that would be quite right for the matter in hand. And he told us too that we would never acquire that art except by long discipline and practice.' As the conversation proceeded he added, half-humorously, half in earnest; 'What a pity some of my army friends haven't learned that lesson better! They issue an order one day; but it is so badly worded that they have to issue another next day to explain or correct it.'

Haig was a logical thinker. Though as a brilliant cavalry leader he could seize the opportunity for swift and audacious decision, in his military planning he showed rather the mind of a brilliant staff officer always. He recoiled from strategical conceptions that were not in accord with military principles, and was impatient of opinions that were not based on facts. He has been criticised for a lack of 'flexibility'; to Wavell, writing in praise of Allenby, Haig seemed to have a 'one-track mind'. But though there may be grounds for the criticism, regard ought to be paid to the conditions in which the war of 1914-18 was fought; and during the first three years there was no obvious display of genius among the leaders on the one side or on the other. In the static conditions of trench warfare the first stern necessity was to contain the enemy, never to relax grip on him, never to allow him opportunity to spring a surprise, and gradually to wear down his strength. Nivelle attempted a brilliant improvization which nearly lost us the war; Haig by his dogged inflexibility prepared the way for final victory. And when in the summer of 1918 the war became at last a war of movement, Haig's gifts as a military thinker were revealed in a new light. Here now he showed a flexibility, a boldness of conception, and a gift of swift and decisive action which amounted almost to genius. And when other Allied leaders, military and civilian, believed that the war must inevitably drag on for another year, Haig, and Haig alone, had the vision to see that by vigorously engaging the enemy a decision might be reached before the winter; and the war was ended by November.

It is much to be regretted that Haig has so often been depicted in prejudiced quarters as dull and unimaginative. Mr Lloyd George has written: 'I never met anyone in a high position who seemed to me so utterly devoid of imagination.' History is not likely to endorse that verdict; it tells us more about Lloyd George than about Haig. Haig certainly never allowed imagination to carry him off his feet; his judgments, though they may at times have been unduly optimistic, were based on a sober assessment of the facts as he saw them. But the stern realist, the logical thinker, was also without doubt a man of vision; and it was this combination of gifts that gave Haig his distinction. Through a close study of the present situation he was always seeing ahead into the needs and possibilities of the future. In the years before 1914 no one in the Services did more than he to ensure that, if the need arose, Britain should have an army (including a 'citizen army' force) organized and equipped for service on the Continent. And when finally war came, no one saw more clearly than he what a long grim struggle lay ahead, or kept more steadily in view the certainty of ultimate victory and the one sure way of attaining it. There was a deep significance in those words which (as I have told[1]) he used to me after the final rout of the German army : 'I always *saw* that it would come to this.'

Haig's serious thinking, from quite early days, had been centred almost entirely on his profession as a soldier. He had acquired a knowledge of French so that he could both speak and understand it. Telling how Haig entertained Clemenceau to lunch in the course of the Somme battle Lord Esher,[2] who was on one of his visits to GHQ, writes : 'Douglas expresses himself perfectly in French, never using any but the exact word he requires.' Haig had not read widely in history or literature. But I always found that, when matters arose which interested him, he brought an independent mind to bear on them. Being a poor conversationalist, he had little taste for verbal argument. But he was ready to listen to a good case presented simply and cogently. And he would sometimes express his appreciation of a sermon by saying that it gave him 'something to think about'. On the other hand, he was quick to detect superficial, slovenly or muddled thinking; and it always got short shrift from him. Verbiage was especially distasteful to him. 'Words! Words! Words!' was a frequent expletive as an expression of criticism or disapproval.

Haig was deeply interested in education; I had many talks with him on the subject. His own interests were not strictly

[1] p. 93.
[2] *Journals and Letters*, iv, 56.

scientific. And he was insistent that, for the training of youth, even of those who hoped to become scientists, something more than scientific education was essential. He once said to me: 'I think of the teachers who have helped me most; and they were not professional scientists.' But this is only one side of the picture. A matter to which he constantly gave close personal attention—one can see this in the pages of his Diary, with their frequent illustrative sketches and detailed descriptions—was the development and production of new weapons, small and great, including the tanks. In his attitude to 'scientific warfare' he was always ready to learn, and far more receptive than some of the commanders in other armies.

Haig's warmth of heart is attested by certain passages in his diary account of the First Battle of Ypres (October-November, 1914) which have remained vividly in my memory since first I read them. (The occasion of reading, I may say, was a Sunday in October, 1917, when something I had said in my sermon at church led Haig and Rawlinson to recall over the lunch-table some of the anxieties of the corresponding period in 1914, one of the most critical periods of the war; and the Chief in his kindly way allowed me to read what he had written in his diary at the time). Amid all the worries of the military situation at that desperate hour in 1914 Haig had been deeply moved by the crowds of refugees that poured down the Menin Road, having thrown away everything they could spare, and 'with a look of absolute terror on their faces, such as I have never before seen on any human being's face'. The diary told also how he had encountered a small party as he was having lunch by the roadside. 'They had walked all the way from Ostend, with a basket on the arm or a pack of clothes on their backs—all that was left to the poor things of their property. I gave them two dozen Oxo soup squares for which they seemed most grateful.' These words were the expression of a full heart. *Sunt lacrimae rerum et mentem mortalia tangunt.*

Haig showed the same sensitive spirit in his concern for the welfare of his own men. There are frequent instances of this in his diary for 1915, when his responsibility was the more immediate one of an army-commander. He was angry when

he found that ('contrary to my orders') men were being kept
too long in water-logged trenches, or through 'fussiness and
over-anxiety' were forbidden to remove their boots; and as a
result he gave divisional and corps commanders 'a good talking
to'. He did not normally interfere so directly when, having
ceased to be an army-commander, he had now as Commander-
in-Chief a number of armies under his command. But even
at GHQ he continued to feel the same personal concern and
responsibility for the men whom he commanded. He had a profound admiration for them as men, based on an appreciative
understanding both of what they had to suffer and what they
succeeded in accomplishing; and the admiration in turn bred
affection. Naturally his human sympathies had to be controlled
by his obligations as a commander. But the sympathy was never
absent. His apparent detachment was not aloofness; still less
was it ruthlessness; it came rather from his joint involvement
with all ranks in a stupendous task, in which he had his own
very special responsibilities.

His understanding and sympathy, which found such devoted
expression in his work on behalf of ex-servicemen and their
dependants, revealed themselves also at times in simple, homely
and even sentimental ways. Copied out in one of his note-
books at Bemersyde were a number of quotations which he
valued as 'mottoes' for daily life; and among these were those
lines of Adam Lindsay Gordon (his favourite quotation, Lady
Haig has called it):

> 'Question not, but live and labour
>   Till the goal is won.
> Helping every feeble neighbour,
>   Seeking help from none.
> Life is mostly froth and bubble,
>   Two things stand like stone;
> Kindness in another's trouble,
>   Courage in your own.'

Kindness was certainly as integral a part of Haig's nature as
courage.

The warmth and generosity of his feelings which he was careful to hide in general conversation would often come to expres-

sion in his personal correspondence. I could give many illustra-
tions of this; I content myself by quoting a letter which has
only recently come to light and is now published here for the
first time. It was written in the course of the First Battle of
Ypres (November 1914) to his nephew, Colonel Oliver Haig
described on the envelope as 'Comt Regt of Sharpshooters'. Here
is the letter :

'near Ypres
6 Nov 14

'My dear Oliver

Heartiest congrats on your promotion to Colonel! I
feel sure you have your regt in excellent order.

We are having pretty hard times here, but are holding our
own.

The enemy's high explosive shell is terrific in force and noise.
They have been bombarding the town here—a nice old place.

You must *not* fret because you are not out here. There is a
great want of troops, and *numbers* are wanted. So I expect you
will *all* soon be in the field. Meantime train your machine guns,
it will repay you.

Forgive hurried line, and with best love I am always your
very affectionate uncle

Douglas.'

The letter takes us back to one of the most critical periods of
the whole war. Fanatical German efforts to force a decision were
directed in the main against Haig's depleted and exhausted
First Army Corps, and more than once it appeared as if nothing
could stop them. Haig himself became a target for their fire;
with the aid of spies direct hits were secured on two chateaux
he had occupied east of Ypres. That such a letter (in his own
handwriting) should have been written at such a time tells us
much of the man who wrote it. Haig's mind was always
concentrated first of all on his duties as a commander; and in
the worst crisis his courage and his calm remained unshaken.
But his military responsibilities never made him indifferent to
his friends; he remained at all times the man of generous
instincts and essential kindliness. Those closing words of greet-

ing, 'with best love' 'your very affectionate uncle' ought to remind us that Haig was certainly not the cold-blooded creature that he has sometimes been made out to be.

He could not have been the man he was without a certain amount of personal ambition; but it was plain that ambition was from a quite early stage combined with an overmastering sense of duty, and was in the end sublimated by it. Duty became one of the watchwords of his life; it made him rigorous in the standards which he set for himself and which he expected from others. Always present to his mind was the service due to King and country. A very early illustration of this is to be found in a letter written in 1902, at the end of the South African war, to the same young nephew Oliver. Oliver Haig did not have the military instincts of his famous uncle; and having served with the 7th Hussars in South Africa he contemplated retiring (on his father's death) to the family estate in Fife. In writing to him on that early occasion Haig 'let himself go' as he very seldom did; he became in it indeed something of a preacher. 'The gist of the whole thing is that I am anxious not only that you should realize your duty to your family, your country and to Scotland, but also to the whole Empire. "Aim high" as the Book says, "perchance ye may attain." Aim at being worthy of the British Empire, and possibly in the evening of your life you may be able to own to yourself that you are fit to settle down in Fife. At present you are not; so be active and busy. Don't let the lives of mediocrities about you deflect you from your determination to belong to the few who can command or guide or benefit our great Empire. Believe me, the reservoir of such men is not boundless. As our Empire grows, so is there a greater demand for them, and it behoves everyone to do his little and try and qualify for as high a position as possible. It is not ambition. This is *duty*.'

Haig had indeed, as one realized in talking with him, a lively appreciation of British history and traditions, of the British way of life, and of the place which Great Britain, and the British Empire, had among the nations of the world. And as he came to see with increasing clearness how our British heritage was under threat, and would almost certainly have to be defended by the sword, so too his faith in the Empire deepened;

and it became a powerful motive force in all his thinking and planning, before, during, and after the war.

He was, however, a reticent man; and deeply as he felt on such matters, he preferred, as a serving soldier, not to voice his feelings in public gatherings, but to go on quietly with the practical tasks assigned to him of bringing the military resources and equipment of Britain, and of the Empire, to the highest possible grade of efficiency. Once the war was over, however, and his military responsibilities had ended, he became more outspoken in proclaiming his patriotic faith. 'Loyalty to King and Empire' was one of the keynotes of his appeal as he went up and down the country on behalf of the British Legion, or travelled overseas to South Africa and Canada to establish the British Empire Service League. And two days before his death, on what was to be his last public appearance, he used these words in addressing a Richmond troop of Boy Scouts, sons of disabled ex-servicemen: 'It is essential that the young should be taught the meaning of Empire and the sacrifices that their fathers have made for it. . . . When you grow up, always remember that you belong to a great Empire, and when people speak disrespectfully of England always stand up and defend your country.'

During the war there was one solemn occasion on which he came near to testifying openly to his confident faith in the destiny of the British Empire. I refer to the occasion of which I have already told,[1] when in one of his darkest hours he sat alone at his desk, writing with his own hand the fateful message which was to tell 'to all ranks' how we stood 'with our backs to the wall' and there was 'no other course open to us but to fight it out'. As the manuscript shows, the message originally closed with the words: 'But be of good cheer, the British Empire must win in the end.' Such indeed was his confident belief. But having written the words he later deleted them; and the Order as finally issued closed without them.

Why did he decide to omit the words? Not, I am sure, because of any lurking fear that a British victory might not after all be possible. More likely he felt, with his realistic outlook,

[1] p. 82.

that the words smacked too much of a facile optimism. The
story is told of a Scottish scholar of an older generation that,
spurning the superficial comfort offered to him at a time of
great distress, he said : 'One steady look into the dark is worth
more than the light of a hundred farthing candles.' Was it
perhaps so with Haig? I like to think that, as he gazed that
morning into the darkness, the conviction came over him that
something more was needed than faith in the British Empire,
that patriotism by itself was not enough, and that, rather than
express confidence in 'things that might be shaken', it was
best that we should let the picture stand as he had painted it
in all its blackness, with the real grounds of his confidence
unexpressed.

Have we perhaps here a reminder that, strong as was Haig's
faith in his country's cause, it had come by this time to be
reinforced by a deeply religious faith, a faith which meant
much for himself in his own soul, but which he would not
lightly invoke in addressing others?

CHAPTER 13

THE MAN OF CHARACTER

His professional qualifications apart, Haig's greatness was a greatness of character; this was emphasized again and again by those who knew him best, both during his command in France and in the years that followed. And it was this, more than anything else, that marked him out among the leading men of his time. Recalling in his book, *Disenchantment*, how a regrettable feature of the post-war years was a set-back in the chivalrous temper of the nation, C. E. Montague goes on to qualify what might seem too sweeping a generalization; and he adds this significant reminder: 'Haig was still alive.' If we ask the secret of the quite unique place which Haig had at that time in the honour of his fellow-countrymen, the answer is that, besides remembering his services as a commander in the field, they now knew him for the sort of man he was, and they gave him their confident and warm-hearted trust.

Character has many facets. Haig's private life was beyond reproach; not a breath of scandal ever sullied his personal reputation. In the language of his time he was 'a perfect gentleman'; no finer gentleman, it was often said, ever wore the King's uniform. Deception and underhand methods were wholly alien to him; some recent attempts to paint a different picture of him (e.g. the allegation that he intrigued with the King to secure the replacement of Sir John French) are based on a completely false reading of the evidence, and it was he himself who (in his Diary) supplied the evidence which his detractors chose to distort.

One essential key to his character was loyalty. As a soldier his first loyalty was to his King and country, and to the cause which he served; and again and again (more especially in his readiness to serve under Foch in 1918, when the part he played in securing Foch's appointment was an active and not a merely passive one) he swept aside purely personal considerations of no importance. How often have I heard him say: 'It matters nothing who gets the credit if only we do what is right.'

H

But no less notable, as all of us knew who had close dealings with him, was his loyalty in personal relationships; and here we have a side of his nature which is apt to be ignored when he is depicted as cold and stiff. He expected men to give of their best; and if they did they could count to the full on his appreciation and support.

It has been finely said of Haig:[1] 'Loyalty was so much of the very fibre of his nature that he could not bring himself to part with anyone to whom he had once given his confidence, unless he was convinced that the individual had proved himself unworthy of it. To throw an unpopular subordinate to the wolves in order to placate public opinion would have appeared to him nothing less than a crime.'

I recall two instances of this, in both of which I knew how deep his feelings were. When pressed by the Secretary for War, towards the end of 1917, to replace his chief Intelligence Officer, Brigadier-General Charteris, he sent in reply a closely-reasoned letter from which I quote these sentences: 'I cannot agree that Charteris should be made "whipping-boy" for the charge of undue optimism brought against myself. . . . If the War Cabinet are not satisfied with the views put forward by me, it is I, and not Charteris, who must answer for those views.' So too, when the Government called for the removal of General Gough after the rout of the Fifth Army in March 1918, Haig was convinced that, with so long a front to defend and with so few reserves, Gough had put up a magnificent fight; and he told the Prime Minister to his face that, if General Gough was to be recalled to England, the Government itself must send an order to that effect.

In assessing the difference of conditions in the two World Wars it is not always remembered that Haig, with five Armies under his command, had other grave responsibilities besides the control of military operations. It was, for example, incumbent on him to work in the closest cooperation with our French allies even to the extent of adjusting his strategical plans to conform with those of the French High Command. Both in

[1] Duff Cooper, *Haig*, ii, 197f.

the political and in the military sphere France as an ally often proved self-centred and exacting; and not the least part of Haig's greatness as a commander came from his success in the difficult field of inter-allied relations. He was punctilious in keeping on the best of terms with his French colleagues, who, apart from the increasing respect which they came to have for his military abilities, valued him for his frankness, fair-mindedness, and complete devotion to the allied cause, and knew that an agreement made with him would be kept. On the other hand Haig could not forget the appalling losses which the French had suffered in 1915 and 1916; and the fear was never far from his mind (and in no small measure it influenced his offensive plans) that our allies might at any time weary of the struggle and leave us in the lurch. In an estimate of Haig's qualities as a man and a commander a factor of profound significance, and one which unfortunately is too often ignored, is his unfailing contribution to allied unity, and his readiness to accept added responsibilities both for himself and for the British army.

It is a notable fact that in the fateful months of 1918 Haig and the new French Premier, M. Clemenceau ('the Tiger') should have drawn so close together, each regarding the other not merely with confidence but with a sincere personal regard. Haig often told me how happy he was in his relations with M. Clemenceau. He had previously had doubts about some other French politicians; but he had none about Clemenceau, and was sure that in the continuance of the fight he would have the full support of the French Premier. His relations too with Foch were founded on a mutual confidence which stood up to every strain. Foch and he were old comrades-in-arms, and they were both essentially 'fighters', ready to co-operate to the full in the common cause. During the period when Foch was Generalissimo Foch certainly owed as much to Haig as Haig did to him. And Foch was generous in acknowledging this. On the day of the official funeral service for Haig in London, and shortly after the departure of his body by train on its last journey northwards, I had gone at her request to call on Lady Haig, and while I was alone with her the maid announced the arrival of Marshal Foch and Marshal Pétain; and in the short talk I had with Foch before I left he expressed his admiration of the British commander in these simple, moving words: *Il était très droit, très sûr, et très gentil.*

Haig's greatness of character was no less marked in another field, his relations with the Home Government. During the greater part of his command most members of the Government appreciated his military ability and gave him their support. It was above all Mr Lloyd George who was lacking in such appreciation, and worked steadily to curtail Haig's powers and (if possible) to secure his removal. It was an extraordinary position for a commander to be placed in—to have under him the greatest army Britain had ever sent to war, to be responsible for the defeat of her most formidable enemy since Napoleon, and yet to be denied the confidence of the Government that kept him at his post. Much will continue to be written on this aspect of the World War I story. I content myself here with some sentences from the pen of Mr Winston Churchill,[1] himself for most of the time a leading member of the Government. After telling how Haig 'at all times treated the Civil Power with respect and loyalty', he goes on to say: 'Even when he knew that his recall was debated among the War Cabinet, he neither sought to marshal the powerful political forces which would have come to his aid, nor failed at any time in faithfulness to the Ministers under whom he was serving. Even in the sharpest disagreement he never threatened resignation when he was strong and they were weak.'

From such a source this is impressive testimony. I cite the following in confirmation of it. Following on the defeat of the Fifth Army in March, 1918, there had been severe criticism of the Prime Minister for keeping troops in Britain that ought to have been available to meet the threatened German attack in France. In reply Mr Bonar Law, with the authority of the Prime Minister, made an important statement in the House of Commons which contained, *inter alia*, the following sentence: 'Notwithstanding the heavy casualties in 1917 the Army in France was considerably stronger on the 1st January, 1918, than on the 1st January, 1917.' Greatly daring, and with a clear

[1] In his collection of essays, *Great Contemporaries*, p. 188.

knowledge that he was thereby sacrificing his career in the Army, Major-General Sir Frederick Maurice, who had been Director of Military Operations at the War Office, wrote a letter to the press categorically denying the truth of the statement. What followed is a complicated story into which I need not enter here;[1] I am concerned rather with Haig's immediate reaction. On hearing that General Maurice had written to the press he wrote that night in his diary : 'This is a grave mistake. No one can be both a soldier and a politician at the same time. We soldiers have to do our duty and keep silent, trusting to Ministers to protect us.'

These closing words are significant. Haig knew that the Prime Minister of the day was very far from being concerned to protect him. But there is nothing cynical in what he wrote. Rather he was giving expression to his own high conception of ministerial honour and responsibility. A soldier is a servant of the Government; the Government appoints him, and if it so decides it has a right to remove him. But so long as he is retained in his command it is the Government's duty to support him. On the other hand it is the soldier's duty to remain at his post, doing the work he has been given to do, and not interfering with the work of Government. For these reasons Haig had no hesitation in regarding Maurice's action, however well-intentioned, as a mistake. Had he been so disposed, he might have read the letter as a welcome defence of the 'soldier' against the criticisms and misrepresentations of the 'politician'. But that is not how he saw the issue. He knew what was incompatible with his duty as a soldier; and any idea that the incident might serve to strengthen his own position would have had no place in his thought.

Previous to this Haig had seen a report of Mr Bonar Law's speech to the Commons, and had been shocked to read in it that the extension of the British line in France was an arrangement agreed to by the British and French Commanders without interference from the British Government. That statement, Haig felt, was inaccurate and misleading; and he accordingly wrote a note which he handed personally at GHQ to Lord Milner,

[1] A detailed revelation of it appeared in 1954 in Lord Beaverbrook's book, *Men and Power*.

Secretary of State for War, giving what he regarded as a true version of the incident. I quote here from his diary of Sunday, April 28th : 'I told Lord Milner I did not wish to embarrass the Government at this time, but I must ask that a true statement of the facts be filed in the War Office. Milner said he would be glad to do this, and that he recollected very well how all along I had objected to any extension. It was his opinion that if we had not taken over some line from the French the blow would have fallen on them and the war would have been well-nigh lost.' It is not surprising that Milner, like most other members of the Government, regarded Haig with increasing appreciation and confidence, and that these two men, however much they differed in other ways, should have found it easy to work together.

In telling my story of Haig I have drawn at many points on his diaries and private papers which over a period of many years I have had freedom to consult as I wished. Haig had early in life acquired the habit of keeping a diary, though like other diarists he did not always keep up the habit. But during the Great War there were few days when he did not find time to write (sometimes quite shortly, at other times in considerable detail) an account of the events and developments which were engaging his special attention. At intervals of a few days a carbon copy of what he had written was sent for safe keeping to Lady Haig in her London home; and I recall the pride with which he once remarked to me : 'No one was ever able to say during the war, Oh ! I got that piece of information from Lady Haig.' After the war the record for each day was typed on a separate sheet; and these sheets, together with a vast collection of official documents, memoranda, letters, etc., have been preserved in thirty-eight massive volumes, which, after being kept for many years at Bemersyde, are now in Edinburgh in the National Library of Scotland.

To the trained historian these diaries and papers are invaluable; no comparable record has been provided by a British commander in any other of our great wars. And a very special interest of Haig's diaries is that they tell, with unaffected frankness, of his relations with our French allies, with the British

Government (and in particular with the Prime Minister), and even with the King. But diaries may become dangerous in the hands of readers who lack either the understanding, or the fairness of mind, to make right use of them. When with the publication in 1952 of Mr Robert Blake's book, *The Private Papers of Douglas Haig*, the public was enabled for the first time to read selections from Haig's papers, the climate of opinion at the time was against the book's having a favourable reception; and Haig's detractors claimed to find in it abundant justification for attacking not merely his military competence but also his character. A picture was presented of him as self-willed, obstinate, ungenerous, disloyal, egotistical, even malicious and a mean intriguer. This is a flagrant misreading of the evidence; and the resultant picture is scandalously false, wholly incompatible with what is otherwise known to be true. It is to be hoped that such misrepresentations will cease.

Whatever may be the verdict of history on Haig's military abilities, there can be no question of the nobility of his character. Many qualities make for success in public life; but none of them is likely to give lasting success where character is wanting. By character is meant that a man's personality, cleansed of all that is mean and unworthy, is unified and directed towards a dominant purpose, and is thereby stabilized and invigorated. And in times of crisis (and war is a crisis from beginning to end) it is character, alike in individuals and in nations, that often proves to be the decisive factor.

'D. H. is behaving wonderfully'—I quote here an entry Lord Esher made in his Journal at that critical period, the end of May, 1918. 'His coolness and detachment of mind under all forms of provocation are admirable, but they are only what I always knew were the qualities he would display. He trained himself in early days in self-control as part of a soldier's equipment. It is the discipline of the mind, and is acquired, and is not innate as many people think. . . . He always gives credit for the best that is in a man, and realizing that all human character is shot-silk he looks at the brighter rather than at the darker colours. It is the saner and safer outlook always. Others are influenced by passion and prejudice, so that they are apt to take wrong turnings. D. H. rarely does, if ever.'

I agree cordially with what Lord Esher has written. I would only add this. Coolness, detachment, self-control were essential qualities in Haig's equipment; but he brought to his task also other qualities of a more positive kind. His convictions strengthened him for action. We are to picture him as

'One who never turned his back but marched breast
                                        forward,
Never doubted clouds would break,
Never dreamed, though right were worsted, wrong would
                                        triumph,
Held we fall to rise, are baffled to fight better,
Sleep to wake.'

Those of us who lived close to Haig never ceased to admire his strength of character—his inner poise which nothing could disturb, his quiet resolution, his readiness to accept responsibility, his power to sink self in the common cause, his invincible faith. Visitors to GHQ, seeing him for the first time, often remarked to me: 'What a burden that man carries! Yet he looks as if he hadn't a care in the world.' Month after month, year after year, amid disappointments and failures, misrepresentations and intrigues, he went steadily on his way, allowing nothing to shake his nerve, to break his spirit, or to sap his confidence in final victory. In some of the most critical periods of the war Haig's indomitable spirit, allied to the indomitable spirit of his troops, provided a bastion against which enemy attacks beat in vain. Finally it provided the vital energy that carried us on to victory.

CHAPTER 14

THE MAN OF FAITH

One naturally hesitates before writing about another man's religion. One hesitates all the more when that man's religion is essentially personal, for then one is treading on holy ground. But no estimate of Haig's character or achievements is likely to be either adequate or trustworthy if it ignores or, what is worse, misinterprets the deep personal religious faith which was so significant a part of his nature in the First World War. Here, however, we are in a field in which it may not be so easy for the student of history in the future to gather his facts, or to see them in their true setting and significance; and anyone whose testimony may be of assistance is under an obligation not to withhold it.

Haig did not wear his religion on his sleeve. Never once during the war, so far as I recall, did he appeal in any public utterance to the Divine Name. On addressing the troops he never invoked the Almighty as an ally. Even in that black hour when he told how we stood with our backs to the wall he gave no indication of a hope that we might look for divine intervention; rather 'each one of us must fight on to the end'.

Something of his personal faith may be learned from a letter which he wrote to Lady Haig a few days before the opening of the Somme battle in 1916. In writing to him she had encouraged him to ask for God's help for the coming offensive, and here are some sentences from his reply. 'Now you must know that every step in my plan has been taken with the Divine help—and I ask daily for aid, not merely in making the plan, but in carrying it out, and this I hope I shall continue to do until the end of all things which concern me on earth. I think it is this Divine help which gives me tranquillity of mind and enables me to carry on without feeling the strain of responsibility to be too excessive. I try to do no more than "do my best and trust in God", because of the reasons I give above. Very many thanks for telling me your views of this side of my work, because it has given me the chance of putting my ideas on paper. For otherwise I would not have written them, as you know I don't talk much on religious subjects.'

If he did not give expression to his religion in public, he showed a similar reticence in private conversation. Had he been disposed to unbosom himself on matters of personal religion, he might presumably have done so to me. Yet he practically never did, either in speech or in writing; the occasions on which he did so might be numbered on the fingers of one hand.

The first occasion was for me so memorable that I recall it here in some detail. It was on the morning of Sunday, March 24, 1918, three days after the opening of the terrific German offensive designed to smash the Allied line and so end the war. I had scarcely expected that in such a situation he would feel free to attend worship as usual, travelling from his Chateau some miles away to our little church on the ramparts of Montreuil. But a few minutes before 9.30 his car appeared; and as he approached to where I awaited him I noted that there was no smile on his face this morning, only a look of calm resolution. He shook hands with me, but said nothing; and so, scarcely knowing what I said, I haltingly expressed the hope that things were not too bad. I did not know that it was a strict rule with the Chief that, when he was on his way to church, none of those who accompanied him, or whom he met, should utter a word in his presence about the war. And so, when he replied tersely that they would never be too bad, I took this to be in part no doubt a gentle rebuke, but in part also an expression of his own calm assurance. It was a solemn moment; and then, as one who understood something of what it all meant for him, I stumbled on and said: 'No! You who were through Mons and Ypres in the first year will never think anything too bad after that.' At once he followed up with a statement which lifted the whole conversation to a higher level; and I recall how I stood transfixed as he uttered it, for I had never before heard him use language that was even remotely like this. 'This,' he said, 'is what you once read to us from Second Chronicles: "Be not afraid nor dismayed by reason of this great multitude; for the battle is not yours, but God's." ' (II Chronicles XX, 15.) And having said that he passed at once into the church.

Here we have what might be termed a confession of faith—a confession all the more significant because it was made in confidence, as from one man to another. And it helps us to see what Haig's religion meant for him as soldier and as Commander-in-Chief. It reveals him as a man essentially humble and devout, he found in his religion a source of inner strength; but it reveals him, too, as a man who never let go his strong grip on reality. By daring to take to himself the Scripture assurance, 'The battle is not yours, but God's', he recognized that, so far from relieving him for a moment of his responsibilities, it rather impelled him to take them the more seriously. The battle might seem indeed to be passing out of human hands; but the responsibility still rested squarely on his shoulders to keep control, so far as possible, of an increasingly critical situation, to plan with due forethought, to take at each step the right decisions—and never to lose heart. We need not wonder that, as his custom was, he came that morning to worship and prayer. And history will tell how that evening he responded to the challenge as he saw it. After a fruitless consultation at a late hour with the French Commander, Marshal Pétain, which revealed the terrifying prospect of the two allied armies being separated, he got back (as I have already related[1]) after midnight to his headquarters at Montreuil, and immediately wired to London urging that steps be taken without delay to secure a Commander with responsibility for the Allied front as a whole. In this way he sacrificed some part of his own freedom of action. But the line was saved.

As a second illustration of Haig's religious faith I may tell of a letter I received from him some three weeks after this, at the time when, after a fresh German onslaught further north, he issued his 'Backs to the Wall' Order. I had not seen him for over a week, for he had spent the previous Sunday in a forward area; and so, when I read what he had to say in that Order, I sent him a personal note. I have found that letter preserved among his private papers, and as a prelude to what he was to write in reply it may be of interest if I quote it.

[1] p. 78.

'GHQ, April 15, 1918.

'As you could not be with us at church yesterday, will you please excuse my dropping you this short note? I welcome the opportunity I usually have on Sunday mornings of wishing you "bon courage", and should just like to do so now by letter. And, what is more to the point than any personal greeting, I would fain send you, in the highest name of all, that benediction I should otherwise have pronounced on you at the close of the service.

'We remembered you in our prayers yesterday—you, and your forces and the great cause you are so greatly defending. There is no man under heaven for whom these days a greater volume of prayer ascends to heaven than for you. May the knowledge of that strengthen you.

'P.S. It goes without saying that this little note needs no answer; and you will pardon, I know, the liberty I take in writing.'

The following day I received this reply from him; it was written throughout in his own hand.

'Tuesday, April 16, 1918.

'My dear Duncan,

One line to thank you most truly for your letter. I am very grateful for your thinking of me at this time, and I *know* I am sustained in my efforts by that Great Unseen Power, otherwise I could not be standing the strain as I am doing.

Yours most truly, D. Haig.

'I missed my Sunday morning greatly. But it could not be helped.'

There is no need to attempt an analysis of Haig's religious faith; I am sure he never analysed it for himself, for he was a man of action, not a theologian. But if we are to appreciate it for what it was, it may not be amiss to distinguish it from what it was not. There was nothing in Haig of that fevered imagination which prompts some servicemen in their religion to ally themselves, for example, with a literal interpretation of Scripture, with a narrow evangelicalism, with so-called spiritualism, or with the tenets of the British-Israel school. His faith was not the product of some fanciful theory; it sprang from a calm recognition of the challenges and needs of everyday life. That being so, it was not a cloak to be put on or off as occasion arose; it was an essential part of the man himself, vitally alert no doubt in times of crisis, but never wholly unrelated to his normal habits of thought and action. I need hardly add that it was very different from that airy optimism which shallow souls all too readily equate with religious faith. If we would understand what Haig's religion meant for him, we must try to see him as he addressed himself to the demands of his profession.

Fundamentally Haig was a stern realist, who brought a trained military intelligence to bear on every situation, and only reached his decisions after a careful weighing-up of the various factors. But to his realism he added an *élan vital* which enabled him, after surveying a situation, to bring all his powers to deal with it, keeping steadily in view the ultimate aim to be achieved and the best and surest way of achieving it. Within its own limits this *élan vital* of the soldier has much in common with what in religious language is called faith, more especially such faith as is extolled in the 'Roll of Honour' in the eleventh chapter of the Epistle to the Hebrews—a faith which takes the very best that a man has to offer and adds to it something which otherwise he might not have, a combination of far-seeing vision, patient endurance, and invincible courage. Haig's faith was essentially practical. There is not the slightest evidence that he ever allowed it to pervert or overrule his military judgment. What it did for him was to give him an unshakeable confidence in victory, a resolute will for victory, and a serenity which remained unclouded in the darkest hour. In war these are invaluable assets; lacking them, the most brilliant commander may at some point 'crack' and fail.

Haig had obviously a deep reverence for the Bible—the result no doubt of his training in boyhood. But it is right not to misinterpret this, and see it, as some recent writers have done, as pointing to a certain abnormality in him. He belonged to a time when the Bible was far more truly a household book than it is today. It was widely read in the home, and the habit of regular church attendance provided the worshipper with a weekly opportunity to hear its message read and expounded. An acquaintance with scripture thus became part of an intelligent person's education. It influenced, perhaps unconsciously, both his ways of speech and his general outlook, without necessarily implying that he knew much about its origins or its theology. I have no reason to think that Haig's knowledge of the Bible was either profound or extensive. He certainly gained from it a sense of the divine Presence and Power; and this assurance meant much to him. In a broad general sense he valued it especially for the deep seriousness that characterized its message from the beginning to the end, and for the light which it shed for him on the whole duty of man. The words from 2nd Chronicles which he quoted to me after the opening of the German offensive in 1918 apparently meant much for him at that time; he had referred to them a month earlier in a letter sent to Lady Haig, in which he said: 'I must say that I feel quite confident, and so do my troops. Personally, I feel in the words of 2nd Chronicles that it is "God's battle"; and I am not dismayed by the numbers of the enemy.' It was not in self-assurance, but in reverence and devoutness, that he dared to take the words of Scripture as applicable to himself, and say: 'I am not dismayed.'

There was one other occasion, and one only, on which he quoted Scripture to me. In a letter[1] written after the war he recalled how helpful it had been to him to have things put 'into proper perspective on the Sundays'; and he went on to say:

'I have a little half sheet of writing paper sent me by some unknown friend, who apologizes for writing to me, saying "People generally jib at Bible quotations. But these are sent in no tract-giving spirit etc."; and he writes:

' "Fear thou not, for I am with thee; be not dismayed" etc. . . . as far as "with the right hand of my righteousness".

"They shall renew their strength, they shall mount up with wings" etc., as far as "they shall walk and not faint".

"Only be thou strong and very courageous."

'That was the whole story; not even a note to say from what

[1] cf. p. 127.

books of the Bible they come. But they are grand passages for
one situated as I was; and curiously enough one or two of your
most striking sermons were preached on the last quotation.'

Haig's attitude to Scripture was certainly devout and
sincere. But it is important not to misjudge it. It was quite
different, for instance, from that of Cromwell. Cromwell was
prone to quoting Biblical texts; and, accepting Scripture as
verbally the Word of God, he was ready on occasion to use it as
a basis for judgment and an authority for action. Not so Haig.
On the very rare, and essentially intimate, occasions on which
he quoted Scripture, it was because, as a spiritually-minded
man, he gained from it illumination and strength. A closer
parallel might be found in Abraham Lincoln, who, when he
had to justify to himself or to others some fateful decision,
would often appeal to some great passage in poetry or in the
Bible. And alike in Haig's case as in Lincoln's it would be
foolish to associate his faith too exclusively with a few inspired
texts. In the one case as in the other it had a foundation far
more solid and secure. But when a man stands alone in a crisis,
it is an invaluable reinforcement to him if, from his knowledge of history or literature, sacred or secular, he can draw on
the experience of others who have struggled and fought and
conquered, and feel assured that in a very real sense he is *not*
alone.

I have often been asked the question: 'How far is it true
that Haig regarded himself as God's appointed agent for the
winning of the war?' Never during the whole time I was with
him did I hear language of that kind from his lips. I am also
certain that, if ever he allowed himself to entertain such a
thought, he did so in all humility, not out of egotism or wishful thinking, but with a sober grasp of the situation as he
saw it. For what was his position, viewed in a strictly matter
of fact way, apart from any religious interpretation that might
be put upon it? Our country was facing the greatest crisis
it had known for a hundred years, a crisis which he had long
foreseen, and for which he had rigorously prepared himself.
Now that the hour had come he found himself, at his country's

call, in a position of supreme responsibility, in which his plans and decisions would go far to determine whether the final issue would be victory or defeat, with all that the issue seemed to imply for the future of our country, for the cause of justice and freedom, and for the progress of humanity. He knew, too, that he had the confidence of his colleagues in the field, and also (despite the distrust of the Prime Minister) the confidence of the country as a whole. Would it be surprising if, in such circumstances, he humbly traced in his position the working of a divine Providence?

It has been said of Abraham Lincoln[1] that his theology was 'limited to an intense belief in a vast and over-ruling Providence. And this Providence, darkly spoken of, was certainly conceived by him as intimately and kindly related to his own life'. In his Presidential candidature Lincoln declared on one occasion, when replying to an ignorant attack made on his religious position: 'I see the storm coming, and I know that His hand is in it. If He has a place and work for me, and I think He has, I believe I am ready. I am nothing, but truth is everything.' Lincoln's thoughts might well have been Haig's, though Haig would probably have kept his thoughts to himself. But to suggest (as has sometimes been done by men who never shared his responsibilities or fathomed the depths of his character) that a sense of a divine mission was allowed by Haig to close his mind or to cloud his judgment is an utter travesty, false in fact and scurrilously unjust to a character of rare nobility. With some men, no doubt, belief in a divine 'call' leads easily to fanaticism. But Haig was no fanatic. There was about him a mental balance which was associated not a little with his stern sense of duty; and like other devout men down the ages he heard in the call of duty the voice of God. He takes his place with those heroic figures (like Moses and Joshua in the Scripture records, or like Cromwell and Lincoln in the story of the nations) who in some critical hour of history begin by recognizing the need for action in the situation which confronts them, and then, in a spirit of obedience and faith in God, find themselves braced to meet it with courage and resolution, and in so doing draw strength from unseen sources.

[1] Lord Charnwood, *Abraham Lincoln*, p. 437.

CHAPTER 15

THE MAN OF FAITH (continued)

I have already told of Haig's weekly attendance at church. His attendance was in no sense a formality; there was clearly a deep seriousness behind it. The question naturally arises: what precisely did church attendance mean for him?

In a letter which he sent me soon after the war, acknowledging photographs I had sent him of the church hut at Montreuil, there is this striking piece of self-revelation. 'You know pretty well,' he wrote, 'what memories the sight of the interior of your little hut will conjure up for me, and what a spirit of peace combined with high resolve came into one on entering it.' Those last words are significant; they are all the more so because Haig, when he took up his pen to write, could generally find the right words to express what he had in mind. *Peace* and *high resolve*. Haig had about him a serenity and a quiet resolution which never seemed to leave him. It was as if he had risen above tensions and fears and the meaner forms of self-interest. Here, apparently, was one respect in which attendance at church meant much for him.

In another post-war letter to me he expressed himself in somewhat different terms. He had allowed me to read a Memorandum on the War (prepared by some members of his Staff) which was to be put away in the British Museum; and in returning it I told him how much new light I had got from it on the very serious difficulties with which he had to contend during the three years of his command. This led him to say in his reply: 'Yes, it was very difficult to keep going *all* the time of the long war. And I am frequently asked now how I managed to do it!' And he added the sentence which I have already quoted[1] to say how much it had meant for him, as Commander-in-Chief, to have things 'put into proper perspective on the Sundays'. Here, from a new angle, we have an impressive revelation of Haig's religious outlook. Peace and

[1] p. 124

I

high resolve are no doubt all-important; but to be maintained they must be based on understanding. Though a soldier by profession, he saw the war as a terrible necessity, forced on the British people and their allies by the might of an aggressive Germany; and repeatedly, in private talks with leading churchmen and others, he urged the importance of keeping steadily before the troops what it was that we were fighting for, that serious moral issues were at stake, and that we were fighting indeed for the cause of humanity. But he had little use for bellicose sermons. And the deeper issues of the war became for him more clear and more compelling when, through his Sunday morning worship, he was helped to see them 'in proper perspective', *sub specie quadam aeternitatis.*

We know now, though personally I knew nothing of this at the time, that in his diary on Sundays Haig almost invariably made reference to the church service; and he frequently added, by way of analysis or comment, a few simple words about the sermon. We do not look in these comments for anything strikingly deep or original; they are essentially the reflections of a man of action. What is impressive about them is not their content, but the fact that they were written at all, and that this man, with his vast and unending responsibilities, should take time to recall quietly some of the lessons which he was carrying forward from his participation in divine service.

To appreciate Haig's religion we must begin by recognizing it as sincerely and unobtrusively personal; when writers who did not know him describe him as 'calvinist' or 'covenanter', they are apt to convey to the reader a wholly false picture. His religion was personal too in another sense: it was not intimately associated with loyalty to any one religious denomination. During all his earlier period in the army he was no doubt 'Church of England': from January 1916 onwards church attendance might have stamped him as 'Church of Scotland' and 'Presbyterian'. But the change (in its early stages at least) was not strictly a change of ecclesiastical allegiance. With his ever-increasing responsibilities he had come to look for spiritual uplift from his participation in Sunday worship. No doubt he had sometimes been disappointed, as he told me he had been on his first Sunday at St Omer after taking over the supreme command. But if, in his new position, he could have felt confident of receiving the spiritual help he so much desired, he would not, I feel sure, have cared greatly what ecclesiastical label the chaplain carried, or whether the service was that of the Church of England or the Church of Scotland. This did not mean, however, that Church divisions were for him a matter of indifference. With his broad and practical outlook he welcomed warmly every evidence of closer cooperation among the Churches and every advance towards greater unity, just as he deprecated strongly any tendency towards what seemed to be exclusiveness or sectarianism.

The Church of England chaplain at Montreuil during the last two years of the war was the Rev. J. N. Bateman Champain, of the Church of St Mary, Redcliffe, Bristol. It was a source of regret to him as to me that GHQ, with its loosely-knit organization, never met for a corporate act of worship. And so in the summer of 1918 (there was no Advanced GHQ that year to break up the unity of Montreuil) we made tentative arrangements to hold on Sunday, August 4th, the anniversary of Britain's entry into the war, a special service of thanksgiving and intercession in which the whole of GHQ could take part. Haig received the proposal with enthusiasm. He had hoped that the Roman Catholics might also join in the service, and was deeply disappointed to find that their church regulations forbade it. My Church of England colleague and I combined with great cordiality in preparing an Order of Service, which was printed; he was to give the sermon, I to take the prayers. On the Sunday morning a congregation of several hundreds, representative of all ranks and of all GHQ units, came together and formed a hollow square in the quadrangle of the Ecole Militaire. It was indeed a memorable service, and there was general satisfaction at GHQ that it had been held.

In our arrangements for the service there was a development which, trifling though it was in itself, is worth recalling for the light which it sheds on Haig's attitude. The Church of England chaplain had suggested to me that he should give over his part in the service to his father-in-law, the Bishop of Kensington, who was to be with him on an official visit to the

army at that time. I offered no objection; but Haig, when he
heard of the proposal, emphatically vetoed it. He had an instinc-
tive dislike (as is shown by repeated references in his diary)
to visitors from civilian life, and not least to churchmen, unless
they came on a highly responsible mission. And in this case
there was a more cogent reason for his veto. This was to be
an official army service; it ought therefore to be conducted by
the official army chaplains, and not by a visitor from outside.
Here he was acting in accord with what had been his normal
attitude in such a situation. Charteris[1] recalls how, when they
were together in India, Haig's practice at Simla had been to
attend the Anglican service; and he adds: 'He never entered
the door of the Scottish Church, though during his term of
office there was a yearly succession of very able Scottish
divines who by their eloquent sermons filled the picturesque
little church to overflowing.'

I have already related[2] how, on that Sunday evening in
July 1917 before the opening of the Passchendaele offensive,
Haig told me of the frank talk he had had at lunch that day
with the Archbishop of York (Dr Cosmo Lang). As Scots by
birth and early training both he and the Archbishop understood well the historical reasons which led Scotland to have its
own National Church, quite distinct from the Church of
England; but the present crisis created a new challenge and a
new opportunity. In his diary for that day Haig gives us his
own account of the talk. 'The Archbishop spoke to me
privately about the necessity for opening the doors of the
Church of England wider. I agreed, and said we ought to aim
at organizing a great Imperial Church to which all honest
citizens of the Empire could belong. In my opinion, Church
and State must advance together, and hold together against
those forces of revolution which threaten to destroy the State.'
Haig's language here is, of course, not meant to be precise, and
the phrase 'all honest citizens of the Empire' must be interpreted in the context. He was clearly thinking of 'good Christian
men', members of one or other of the Christian communions in
Great Britain or throughout the Empire, whose religious life

[1] *Haig*, p. 60.
[2] p. 60.

might find fuller and more effective expression within the larger framework of what he called an Imperial Church.

That out of the stresses of war there should be born one strong united National Church, exercising influence and authority for the whole of Britain and indirectly for the Empire —this was for him more than a pleasing fancy, a theoretical ideal; it seemed to be a practical necessity without which the nation, and the Empire, might find it hard to hold together. He made a strong plea in support of this in an interview with the King in April 1919. He did so again when he came in May as an honoured guest to the General Assembly of the Church of Scotland. And he pled for it with even greater urgency and insistence when a few days later in Edinburgh he addressed a meeting of Chaplains of various denominations who had served in the war. 'The time for speaking,' he said, 'has gone; the moment has come for deeds. If we are in earnest to get something done, we must be content to give the executive power to a few, and to work loyally and with all the energy that is in us to carry out their directions.' If he were alive today, it would doubtless disappoint him to see how far even now his dream has fallen short of fulfilment, though something has been gained by the establishment of the British Council of Churches and the World Council of Churches and by the spread in countless ways of the spirit of unity.

His devotion to this ideal continued to the end. The last letter that I had from him, written in the month of his death, revealed his deep distress that difficulties over the question of 'intercommunion' should have asserted themselves in a well-known fellowship of ex-servicemen. But I recall more especially, as a tribute to the breadth and generosity of his outlook, the attitude he displayed when in 1927 the Church of England, seeking Parliamentary sanction for the adoption of a Revised Prayer Book, encountered serious opposition from certain sections of opinion in both England and Scotland. On that occasion he made a special journey to London, listened patiently as a member of the House of Lords to the arguments put forward on one side and on the other, and finally gave his vote in favour of the Revised Book, being convinced that, if the Church of England wished to carry out a much needed reform, it would be in the best interests of both Church and State that it should be left free to do so.

Apart from the faithful performance of his duties as an elder in the little parish church near his Scottish home at Bemersyde, this was perhaps Haig's last official action in regard to church affairs. He gave his vote that day in accordance with his sense of duty as a patriotic citizen and a churchman. He had no great grasp of ecclesiastical questions, and had indeed little interest in them. We ought to think of him primarily as a man with a simply personal religion, a religion which profoundly affected his whole outlook on life and his relations with his fellows. While it made him resolute and strong, it made him also sympathetic and tolerant, essentially broad-minded and forward-looking. In the post-war years he had no hesitation in identifying himself humbly and sincerely with the fellowship, worship and work of the Church of Scotland. But his churchmanship was not moulded in accordance with any rigid ecclesiastical pattern. It was bound up rather with his personal sense of duty—with his devotion to his country's welfare, to the needs of his fellow-men, and to the service of humanity in the widest sense. As he said in that address to Chaplains from which I have already quoted: 'What is above all necessary is the spread of brotherhood and comradeship among men and peoples to prevent the re-occurrence of world wars; and that is preeminently the task of the Churches.'

CHAPTER 16

EPILOGUE

The words printed at the opening of this book are a translation of those with which the Roman historian Tacitus opens
his account of the life of his father-in-law, Agricola, the
distinguished military commander who carried the Roman
conquest of Britain to its northern limit in Caledonia. Expressed
in quite general terms his plea was that greatness, when it
appears, should not go unrecognized and unrecorded, and that,
despite the sinister forces that tend to breed depreciation and
forgetfulness, posterity should never be allowed to ignore its
debt to the great men of the past.

I have had one aim in writing the foregoing pages: it is to
tell of Haig as I knew him, in the hope that what I have
written may help others to appreciate his character and assess
his achievements. There have been few men in recent generations who in their lifetime were more honoured by those who
were most intimately associated with them; few, too, who since
their death have suffered more from those twin forces against
which Tacitus issued a warning, *ignorantia recti et invidia*.
A false legend has been allowed to grow up in which essential
facts about Haig are ignored or distorted; and it may be said
of some of the more ungenerous or denunciatory presentations
of him which have appeared in print that they could only have
come from men who were not privileged to know him, or whose
own standards of judgment and behaviour rendered them
incapable of recognizing and appreciating his greatness.

Some of the more adversely critical studies of Haig's
command have gained a certain measure of credibility because
of the apparent fulness of their documentation; they are based,
it is claimed, on a careful and comprehensive study of first-
hand information or of official documents. But, leaving out of
account the question whether a still more careful and comprehensive study may not in time lead to different conclusions,
we do well to remind ourselves that the most careful study of
'facts' may be vitiated if the standpoint from which they are

studied is prejudiced or false. In contrast to the so-called
'scientific' temper in which historians have sometimes thought
it appropriate to approach their subject, the historian of
today is generally disposed to recognize that his business is
not with facts alone, but with facts seen in their true light and
properly interpreted. He must try to enter into the thoughts and
feelings of the men of whom he writes; and with regard to
events he must consider not merely what meaning he himself
in retrospect attaches to them, but what meaning they had for
those who had a part in them and helped to shape them. 'The
art of the historian,' it has been said,[1] 'finds its closest affinity
in that of the portrait painter, who has the double task of
producing a good likeness and a work of art. . . . Like the
portrait painter, the historian must achieve not merely
accuracy of detail, but truth of general aspect and of propor-
tion. This latter indeed is the more important of the two; and
a historian who is inaccurate in detail but correct in the broad
view is less misleading than the one who is accurate but mis-
taken.'

Criticisms of Haig began to appear in his lifetime, but he
preferred to let them go unanswered. Mr Lloyd George, at the
peak of his glory as the victorious Prime Minister, was violent
in his attacks on a military leader whose war-time policy had
run counter to his own. After Haig's death he issued, in six
volumes, his *War Memoirs*. Coming from such a source this
ought to have been a work of first-class historical value; but
unfortunately the spirit in which it was written deprives it of
any such claim. Of Haig he writes with malice and rancour, not
merely claiming to show him up as a military mediocrity, but
accusing him of failures of character and conduct that amount
to criminality.

Unfortunately Mr Lloyd George's work left a legacy behind
it. It came at a time when many in Britain, haunted by
memories of the losses incurred in the war and of the devasta-
tion that had followed, were prepared to look back on the
whole experience with a sense of horror, to analyse the causes
of it, to deal critically with its failures, and to look about for

[1] R. E. Balfour, Essay on *History* in a volume entitled *Cambridge University Studies*, p. 197.

someone to bear the blame. Here many found it easy to follow Mr Lloyd George's vociferous lead; and confining their outlook solely to British losses, they yielded to the belief that these were attributable in the main to military incompetence, and that the war-time hero, Douglas Haig, was after all a man of straw.

The British public thus became the victims of a ghastly *trauma*—the sort of 'strong delusion' which, as St Paul[1] recognized in his day, sometimes overtakes men and makes them 'believe a lie'. Such a delusion while it lasts may do incalculable harm, and it may not easily be dispelled. Yet in time the facts will speak for themselves. And one may trust the British people's sense of fairness; for all its occasional gullibility it has a shrewd instinct in its choice of those whom it is disposed to honour; and more particularly in the realm of character it does not as a rule fail to recognize true greatness when it sees it.

Haig's image grows in impressiveness as we come to appreciate better the 1914-18 war situation. This was the greatest land war in British history. We had not faced anything comparable to it for one hundred years. Both in the scale of its operations and in the complexity of its problems it went far beyond the experience of any of those who took part in it; and this was true of allied and enemy leaders no less than our own. If our commanders were constantly baffled, and had to proceed by a costly system of trial and error, that need not surprise us. What may well cause wonder and evoke admiration is the fact that they emerged victorious, and did so in the space of little more than four years. Misrepresentations that have long been prevalent about the conduct of the war are now giving way before more patient, fair minded and enlightened historical study. The protracted and costly battles of 1916 and 1917 are (as we now see) not to be considered in isolation, as monuments to military incapacity and useless bloodshed; they must be seen as an unavoidable prelude to the victory that followed so dramatically in 1918. And it is wholly at variance

[1] II Thessalonians, ii, 11.

with the facts of the situation to follow Mr Lloyd George in connecting Haig's name solely with the defeats, and giving all the glory of victory to Foch. Nineteen-eighteen was the year of Haig's vindication; and while all due honour must go to Foch for his part in bolstering up the French and keeping the Allied line intact, Haig was in that critical year the main directing genius both in the work of defence and in the final victorious assaults. More than any other military leader he deserves to be recognized as 'the architect of victory'.

Criticisms of the First World War have too often reflected the hindsight of a later generation; there has been a tendency to forget that that war had to be fought without the experience and training, the weapons and equipment, that were available to commanders in the second war. What is in many ways a more instructive parallel may be found in the civil war that had rocked America fifty years earlier. That war too had to be fought by hastily raised levies of civilian soldiers; both North and South suffered staggering casualties; and the campaign was drawn out for four years before victory came at last to the North. Not suprisingly many of the generals on both sides lacked the military and the moral qualities that make for victory. But on each side there emerged a supreme commander who in retrospect still holds the honour and the admiration of his fellow countrymen, and of many also in other lands; and with both Grant and Lee, though in different ways, Haig is not unworthy to stand comparison.

In personal character—his austerity and reserve, his nobility of outlook and behaviour, his rigid adherence to principle, his quiet self-confidence firmly rooted in his trust in God—Haig takes his place alongside that great soldier of the South, Robert E. Lee. But in the military policy which he pursued with unwavering determination his prototype was Ulysses S. Grant. While other Federal generals were hesitant and bewildered, Grant saw clearly that if the formidable Confederate Army was to be defeated, he must hold it in an inexorable grip. He must follow it wherever it went, attack it even in strongly entrenched positions, and he must never allow it time to rest or recuperate. His losses were inevitably heavy; and for that reason he was accused, as Haig was, of incapacity and callousness. So far from being callous Grant had, like Haig, a singular delicacy of feeling, and he was deeply sensitive to the suffering and loss which his policy entailed. But as a commander the one thought that possessed his soul was the necessity to bring the war to a victorious end. He was convinced that this could be done, and could only be done, by steadily undermining the enemy's will to fight, and inflicting on them losses which could not be replaced. And the event proved him to be right.

In the prosecution of his strategical aim Grant had the full support of his President; for Lincoln had the same clear perception of essentials, and pursued them with the same persistence. Haig was not so fortunate. With a Prime Minister who showed no confidence either in his ability or in his policy, Haig had to go his way alone, bringing to his task not merely the military capacity of a Grant but the moral stability and strength of a Lincoln. He might have said with Lincoln : 'We accepted this war; we did not begin it. We accepted it for an object, and when that object is accomplished the war will end.' So too he might have said with Lincoln : 'I do the very best I know how, the very best I can; and I mean doing so until the end. If the end brings me out all right, what is said about me won't amount to anything. If the end brings me out wrong, then angels swearing I was right would make no difference.' And, recalling as we must how Haig's work in the war found a natural culmination in his devoted service after the war was over, we can picture him making his own those other words of Lincoln's in his Second Inaugural Address : 'Let us strive on to finish the work we are in; to bind up the nation's wounds, to care for him who shall have borne the battle and for his widow and orphan, to do all which may achieve and cherish a just and lasting peace among ourselves, and with all nations.'

Two world-wars, with all the losses and suffering involved, have created in the minds of many such a hatred of war that they recoil from every thought of it. One can understand their reaction. Yet there are aspects of war which redeem it from the ruthlessness and brutality so often associated with it. By a striking paradox multitudes of simple men and women rise in war to heights that shame the lower levels on which so many live in times of peace. And while this is true of the common soldier, who after all is the real hero in war, how often it is true also of those exceptional men who are called to positions of high command. Their decisions may involve not merely the lives of thousands but the whole issue of victory or defeat; yet they go about their task with quiet, purposeful devotion, not greedy of quick returns or of personal gain, but keeping their eye steadily on the ultimate goal and prepared if need be in a crisis to stand alone. And the greatest of our commanders, so far from allowing their profession to 'brutalize' or degrade them, seem rather to be purified and ennobled by it. In the second of our world-wars John Dill, Alan Brooke, Archibald Wavell, Andrew Cunningham (to mention four who have now reached the end of their earthly war) will on the score of personal character alone go down in history as among the noblest figures of our generation.

Among our military leaders in World War I Douglas Haig has a place of honour all his own. He has not escaped criticism. But a common saying of his was: 'I don't mind criticism so long as people get the facts right and are honestly concerned for the truth.' He needs no rehabilitation in the esteem of his countrymen. What is necessary is that his countrymen should, in his phrase, 'get the facts right,' and come to know him for the man that he was. And when they set themselves to live over again the story of the First World War, seeing it as an epic whole in which the tragic failures and losses of those four years of war led on to the dramatic climax of November, 1918, they will see emerging in vivid clear perspective the dominating figure of the British Commander-in-Chief. Well indeed it was for the British nation that in one of the most critical periods in its history it had ready at call as the leader of its armies a man with the military gifts, the sublime faith, and the unconquerable soul of Douglas Haig.

INDEX